ASSESSING THE RELIABILITY OF COMPLEX MODELS

MATHEMATICAL AND STATISTICAL FOUNDATIONS OF VERIFICATION, VALIDATION, AND UNCERTAINTY QUANTIFICATION

Committee on Mathematical Foundations of Verification, Validation, and Uncertainty Quantification

Board on Mathematical Sciences and Their Applications

Division on Engineering and Physical Sciences

NATIONAL RESEARCH COUNCIL
OF THE NATIONAL ACADEMIES

THE NATIONAL ACADEMIES PRESS
Washington, D.C.
www.nap.edu

THE NATIONAL ACADEMIES PRESS   500 Fifth Street, NW   Washington, DC 20001

NOTICE: The project that is the subject of this report was approved by the Governing Board of the National Research Council, whose members are drawn from the councils of the National Academy of Sciences, the National Academy of Engineering, and the Institute of Medicine. The members of the committee responsible for the report were chosen for their special competences and with regard for appropriate balance.

This project was supported by the Department of Energy under Contracts DE-AT01-07NA78285 and DE-DT0001857, Task Order 9, and by the Air Force Office of Scientific Research under Contract FA9550-10-1-0435. Any opinions, findings, conclusions, or recommendations expressed in this publication are those of the author(s) and do not necessarily reflect the views of the organizations or agencies that provided support for the project.

Cover image credit: The galaxy image at the bottom is of white dwarf stars in globular cluster NGC 6397—Hubble Space Telescope. Courtesy of NASA, ESA, and H. Riches, University of British Columbia.

International Standard Book Number-13: 978-0-309-25634-6
International Standard Book Number-10: 0-309-25634-8

Additional copies of this report are available from the National Academies Press, 500 Fifth Street, NW, Keck 360, Washington, DC 20001; (800) 624-6242 or (202) 334-3313; http://www.nap.edu.

Suggested citation: National Research Council. 2012. Assessing the Reliability of Complex Models: Mathematical and Statistical Foundations of Verification, Validation, and Uncertainty Quantification. Washington, D.C.: The National Academies Press.

Copyright 2012 by the National Academy of Sciences. All rights reserved.

Printed in the United States of America

THE NATIONAL ACADEMIES
Advisers to the Nation on Science, Engineering, and Medicine

The **National Academy of Sciences** is a private, nonprofit, self-perpetuating society of distinguished scholars engaged in scientific and engineering research, dedicated to the furtherance of science and technology and to their use for the general welfare. Upon the authority of the charter granted to it by the Congress in 1863, the Academy has a mandate that requires it to advise the federal government on scientific and technical matters. Dr. Ralph J. Cicerone is president of the National Academy of Sciences.

The **National Academy of Engineering** was established in 1964, under the charter of the National Academy of Sciences, as a parallel organization of outstanding engineers. It is autonomous in its administration and in the selection of its members, sharing with the National Academy of Sciences the responsibility for advising the federal government. The National Academy of Engineering also sponsors engineering programs aimed at meeting national needs, encourages education and research, and recognizes the superior achievements of engineers. Dr. Charles M. Vest is president of the National Academy of Engineering.

The **Institute of Medicine** was established in 1970 by the National Academy of Sciences to secure the services of eminent members of appropriate professions in the examination of policy matters pertaining to the health of the public. The Institute acts under the responsibility given to the National Academy of Sciences by its congressional charter to be an adviser to the federal government and, upon its own initiative, to identify issues of medical care, research, and education. Dr. Harvey V. Fineberg is president of the Institute of Medicine.

The **National Research Council** was organized by the National Academy of Sciences in 1916 to associate the broad community of science and technology with the Academy's purposes of furthering knowledge and advising the federal government. Functioning in accordance with general policies determined by the Academy, the Council has become the principal operating agency of both the National Academy of Sciences and the National Academy of Engineering in providing services to the government, the public, and the scientific and engineering communities. The Council is administered jointly by both Academies and the Institute of Medicine. Dr. Ralph J. Cicerone and Dr. Charles M. Vest are chair and vice chair, respectively, of the National Research Council.

www.national-academies.org

COMMITTEE ON MATHEMATICAL FOUNDATIONS OF VERIFICATION, VALIDATION, AND UNCERTAINTY QUANTIFICATION

MARVIN L. ADAMS, Texas A&M University, *Co-Chair*
DAVID M. HIGDON, Los Alamos National Laboratory, *Co-Chair*
JAMES O. BERGER, Duke University
DEREK BINGHAM, Simon Fraser University
WEI CHEN, Northwestern University
ROGER GHANEM, University of Southern California
OMAR GHATTAS, University of Texas at Austin
JUAN MEZA, University of California, Merced
ERIC MICHIELSSEN, University of Michigan
VIJAYAN N. NAIR, University of Michigan
CHARLES W. NAKHLEH, Sandia National Laboratories
DOUGLAS NYCHKA, National Center for Atmospheric Research
STEPHEN M. POLLOCK, University of Michigan (retired)
HOWARD A. STONE, Princeton University
ALYSON G. WILSON, Institute for Defense Analyses
MICHAEL R. ZIKA, Lawrence Livermore National Laboratory

Staff

NEAL GLASSMAN, Study Director
MICHELLE SCHWALBE, Associate Program Officer
BARBARA WRIGHT, Administrative Assistant

BOARD ON MATHEMATICAL SCIENCES AND THEIR APPLICATIONS

C. DAVID LEVERMORE, University of Maryland, *Chair*
TANYA STYBLO BEDER, SBCC, Inc.
PATRICIA FLATLEY BRENNAN, University of Wisconsin-Madison
GERALD G. BROWN, U.S. Naval Postgraduate School
LOUIS ANTHONY COX, JR., Cox Associates
BRENDA L. DIETRICH, IBM T.J. Watson Research Center
CONSTANTINE GATSONIS, Brown University
DARRYLL HENDRICKS, UBS Investment Bank
KENNETH L. JUDD, The Hoover Institution
DAVID MAIER, Portland State University
JAMES C. McWILLIAMS, University of California, Los Angeles
JUAN MEZA, University of California, Merced
JOHN W. MORGAN, Stony Brook University
VIJAYAN N. NAIR, University of Michigan
CLAUDIA NEUHAUSER, University of Minnesota, Rochester
J. TINSLEY ODEN, University of Texas at Austin
DONALD G. SAARI, University of California, Irvine
J.B. SILVERS, Case Western Reserve University
GEORGE SUGIHARA, University of California, San Diego
EVA TARDOS, Cornell University
KAREN VOGTMANN, Cornell University
BIN YU, University of California, Berkeley

Staff

SCOTT WEIDMAN, Director
NEAL GLASSMAN, Senior Program Officer
MICHELLE SCHWALBE, Associate Program Officer
BARBARA WRIGHT, Administrative Assistant

Acknowledgments

This report has been reviewed in draft form by individuals chosen for their diverse perspectives and technical expertise, in accordance with procedures approved by the National Research Council's Report Review Committee. The purpose of this independent review is to provide candid and critical comments that will assist the institution in making its published report as sound as possible and to ensure that the report meets institutional standards for objectivity, evidence, and responsiveness to the study charge. The review comments and draft manuscript remain confidential to protect the integrity of the deliberative process. We wish to thank the following individuals for their review of this report:

Andrew Booker, Boeing Corporation
Donald Haynes, Los Alamos National Laboratory
Max Morris, Iowa State University
William Oberkampf, WLO Consulting
J. Tinsley Oden, University of Texas at Austin
Elaine Oren, Naval Research Laboratory
Naomi Oreskes, University of California, San Diego
Susan Sanchez, Naval Postgraduate School
Kaspar Willam, University of Colorado at Boulder

Although the reviewers listed above have provided many constructive comments and suggestions, they were not asked to endorse the conclusions or recommendations nor did they see the final draft of the report before its release. The review of this report was overseen by Ali Mosleh. Appointed by the National Research Council, he was responsible for making certain that an independent examination of this report was carried out in accordance with institutional procedures and that all review comments were carefully considered. Responsibility for the final content of this report rests entirely with the authoring committee and the institution.

The committee also acknowledges the valuable contribution of the following individuals, who provided input at the meetings on which this report is based:

Mark Anderson, Los Alamos National Laboratory
Bilal Ayyub, University of Maryland

Chris Barrett, Virginia Tech
Louis Anthony Cox, Cox Associates
Andrew Dienstfrey, National Institute of Standards and Technology
Michael Eldred, Sandia National Laboratories
Peter Gleckler, Lawrence Livermore National Laboratory
Thuc Hoang, National Nuclear Security Administration
Richard Klein, Lawrence Livermore National Laboratory
Alex Levis, George Mason University
J. Tinsley Oden, University of Texas at Austin
Mikel Petty, University of Alabama in Huntsville
Bruce Robinson, Los Alamos National Laboratory
Susan Sanchez, Naval Postgraduate School
Christopher Sims, Princeton University
Timothy Trucano, Sandia National Laboratories

The co-chairs also thank the following individuals for their helpful discussions over the course of this study:

Donald Estep, Colorado State University
Elizabeth Keating, Los Alamos National Laboratory
James McWilliams, University of California, Los Angeles
Robert Moser, University of Texas at Austin
Habib Najm, Sandia National Laboratories
Leonard Smith, London School of Economics
Karen Willcox, Massachusetts Institute of Technology

Contents

SUMMARY   1

1   INTRODUCTION   7
    1.1   Overview and Study Charter, 7
    1.2   VVUQ Definitions, 8
    1.3   Scope of This Study, 9
        1.3.1   Focus on Prediction with Physics/Engineering Models, 9
        1.3.2   Focus on Mathematical and Quantitative Issues, 9
    1.4   VVUQ Processes and Principles, 10
        1.4.1   Verification, 10
        1.4.2   Validation, 11
        1.4.3   Prediction, 11
        1.4.4   Uncertainty Quantification, 12
        1.4.5   Key VVUQ Principles, 13
    1.5   Uncertainty and Probability, 13
    1.6   Ball-Drop Case Study, 14
        1.6.1   The Physical System, 16
        1.6.2   The Model, 16
        1.6.3   Verification, 16
        1.6.4   Sources of Uncertainty, 16
        1.6.5   Propagation of Input Uncertainties, 17
        1.6.6   Validation and Prediction, 17
        1.6.7   Making Decisions, 17
    1.7   Organization of This Report, 18
    1.8   References, 18

2   SOURCES OF UNCERTAINTY AND ERROR   19
    2.1   Introduction, 19
    2.2   Projectile-Impact Example Problem, 20

2.3 Initial Conditions, 23
2.4 Level of Fidelity, 24
2.5 Numerical Accuracy, 24
2.6 Multiscale Phenomena, 25
2.7 Parametric Settings, 26
2.8 Choosing a Model Form, 26
2.9 Summary, 26
2.10 Climate-Modeling Case Study, 27
    2.10.1 Is Formal UQ Possible for Truly Complex Models?, 28
    2.10.2 Future Directions for Research and Teaching Involving UQ for Climate Models, 29
2.11 References, 30

3 VERIFICATION

3.1 Introduction, 31
3.2 Code Verification, 32
3.3 Solution Verification, 33
3.4 Summary of Verification Principles, 35
3.5 References, 36

4 EMULATION, REDUCED-ORDER MODELING, AND FORWARD PROPAGATION

4.1 Approximating the Computational Model, 38
    4.1.1 Computer Model Emulation, 38
    4.1.2 Reduced-Order Models, 39
4.2 Forward Propagation of Input Uncertainty, 41
4.3 Sensitivity Analysis, 42
    4.3.1 Global Sensitivity Analysis, 43
    4.3.2 Local Sensitivity Analysis, 44
4.4 Choosing Input Settings for Ensembles of Computer Runs, 46
4.5 Electromagnetic Interference in a Tire Pressure Sensor: Case Study, 46
    4.5.1 Background, 46
    4.5.2 The Computer Model, 46
    4.5.3 Robust Emulators, 48
    4.5.4 Representative Result, 49
4.6 References, 49

5 MODEL VALIDATION AND PREDICTION

5.1 Introduction, 52
    5.1.1 Note Regarding Methodology, 54
    5.1.2 The Ball-Drop Example Revisited, 57
    5.1.3 Model Validation Statement, 58
5.2 Uncertainties in Physical Measurements, 59
5.3 Model Calibration and Inverse Problems, 60
5.4 Model Discrepancy, 63
5.5 Assessing the Quality of Predictions, 67
5.6 Automobile Suspension Systems Case Study, 70
    5.6.1 Background, 70
    5.6.2 The Computer Model, 70
    5.6.3 The Process Being Modeled and Data, 70
    5.6.4 Modeling the Uncertainties, 71
    5.6.5 Analysis and Results, 72
5.7 Inference from Multiple Computer Models, 74

CONTENTS   xi

    5.8 Exploiting Multiple Sources of Physical Observations, 75
    5.9 PECOS Case Study, 75
        5.9.1 Overview, 75
        5.9.2 Verification, 76
        5.9.3 Code Verification, 76
        5.9.4 Solution Verification, 77
        5.9.5 Validation, 78
    5.10 Rare, High-Consequence Events, 79
    5.11 Conclusion, 80
    5.12 References, 83

6 MAKING DECISIONS   86
    6.1 Overview, 86
    6.2 Decisions Within VVUQ Activities, 86
    6.3 Decisions Based on VVUQ Information, 87
    6.4 Decision Making Informed by VVUQ in the Stockpile Stewardship Program, 88
    6.5 Decision Making Informed by VVUQ at the Nevada National Security Site, 90
        6.5.1 Background, 90
        6.5.2 The Physical System, 91
        6.5.3 Computational Modeling of the Physical System, 92
        6.5.4 Parameter Estimation, 92
        6.5.5 Making (Extrapolative) Predictions and Describing Uncertainty, 93
        6.5.6 Reporting Results to Decision Makers and Stakeholders, 93
    6.6 Summary, 93
    6.7 References, 94

7 NEXT STEPS IN PRACTICE, RESEARCH, AND EDUCATION FOR VERIFICATION,   95
    VALIDATION, AND UNCERTAINTY QUANTIFICATION
    7.1 VVUQ Principles and Best Practices, 95
        7.1.1 Verification Principles and Best Practices, 96
        7.1.2 Validation and Prediction Principles and Best Practices, 97
    7.2 Principles and Best Practices in Related Areas, 98
        7.2.1 Transparency and Reporting, 98
        7.2.2 Decision Making, 99
        7.2.3 Software, Tools, and Repositories, 99
    7.3 Research for Improved Mathematical Foundations, 100
        7.3.1 Verification Research, 100
        7.3.2 UQ Research, 101
        7.3.3 Validation and Prediction Research, 102
    7.4 Education Changes for the Effective Integration of VVUQ, 103
        7.4.1 VVUQ at the University, 103
        7.4.2 Spreading the Word, 106
    7.5 Closing Remarks, 106
    7.6 References, 106

APPENDIXES

A Glossary   109
B Agendas of Committee Meetings   120
C Committee Biographies   124
D Acronyms   130

Summary

Computational models that simulate real-world physical processes are playing an ever-increasing role in engineering and physical sciences. These models, encoding physical rules and principles such as Maxwell's equations or the conservation of mass, are typically based on differential and/or integral equations. Advances in computing hardware and algorithms have dramatically improved the ability to computationally simulate complex processes, enabling simulation and analysis of phenomena that in the past could be addressed only by resource-intensive experimentation, if at all.

Computational models are being used to study processes as large scale as the evolution of the universe and as small scale as protein folding. They are used to predict the future state of Earth's climate and to decide among alternative product designs in manufacturing. Nevertheless, regardless of their underlying mathematical formalism or their intended purpose, they share a common feature—they are not reality.

Models differ from reality for a variety of reasons. Key model inputs—initial conditions, boundary conditions, or important parameters controlling the model—are usually not known with certainty or are inadequately described. For example, an ocean model must be initialized with temperature, salinity, pressure, velocity, and so on over the entire planet before it can run, but these variables are not precisely known. Another source of discrepancy between model and reality is the approximations that are necessary for representing mathematical concepts within a computational model. For example, the ocean must be represented on a grid, or some other finite data structure, and computational operations propagating this ocean over time are only approximations of mathematics defined on the continuum. More fundamentally still, models deviate from reality because they necessarily ignore some phenomena and represent others as simpler than they really are. Without such omissions and simplifications the models would be intractably complicated.

Given inevitable flaws and uncertainties, how should computational results be viewed by those who wish to act on them? The appropriate level of confidence in the results must stem from an understanding of a model's limitations and the uncertainties inherent in its predictions. Ideally this understanding can be obtained from three interrelated processes that answer key questions:

- *Verification.* How accurately does the computation solve the underlying equations of the model for the quantities of interest?
- *Validation.* How accurately does the model represent reality for the quantities of interest?

- *Uncertainty quantification* (UQ). How do the various sources of error and uncertainty feed into uncertainty in the model-based prediction of the quantities of interest?

Computational scientists and engineers have made significant progress in developing these processes and using them to produce not just a single predicted value of a physical quantity of interest (QOI) but also information about the range of values that the QOI may have in light of the uncertainties and errors inherent in a computational model. However, there remain many open questions, including questions about the mathematical foundations on which various processes and methods are based or could be based.

In recognition of the importance of computational simulations and the need to understand uncertainties in their results, the Department of Energy's (DOE's) National Nuclear Security Administration, the DOE's Office of Science, and the Air Force Office of Scientific Research requested that the National Research Council study the mathematical sciences foundations of verification, validation, and uncertainty quantification (VVUQ) and recommend steps that will lead to improvements in VVUQ capabilities. The statement of task is as follows:

- A committee of the National Research Council will examine practices for VVUQ of large-scale computational simulations in several research communities.
- The committee will identify common concepts, terms, approaches, tools, and best practices of VVUQ.
- The committee will identify mathematical sciences research needed to establish a foundation for building a science of verification and validation (V&V) and for improving the practice of VVUQ.
- The committee will recommend educational changes needed in the mathematical sciences community and mathematical sciences education needed by other scientific communities to most effectively use VVUQ.

KEY PRINCIPLES AND PRACTICES

The Committee on Mathematical Foundations of Verification, Validation, and Uncertainty Quantification views its charge as emphasizing the mathematical aspects of VVUQ and, because of the breadth of the subject overall, has limited it focus to physics-based and engineering models. However, much of its discussion applies more broadly. Although the case studies presented in this report include physics or engineering considerations, they are meant to illuminate mathematical aspects of the associated VVUQ analysis. The committee noted several key VVUQ principles: As a first step toward identifying best practices,

- VVUQ tasks are interrelated. A solution-verification study may incorrectly characterize the accuracy of a code's solution if code verification was inadequate. A validation assessment depends on the assessment of numerical error produced by solution verification and on the propagation of model-input uncertainties to computed QOIs.
- The processes of VVUQ should be applied in the context of an identified set of QOIs. A model may provide an excellent approximation to one QOI in a given problem while providing poor approximations to other QOIs. Thus, the questions that VVUQ must address are not well posed unless the QOIs have been defined.
- Verification and validation are not yes/no questions with yes/no answers, but rather are quantitative assessments of differences. Solution verification characterizes the difference between a computational model's solution and that of the underlying mathematical model. Validation involves quantitative characterization of the difference between computed QOIs and true physical QOIs.

Specific to verification, the committee identified several guiding principles and associated best practices. The main text discusses all of these and provides supporting detail. Some of the more important principles and practices are summarized here:

- Principle: Solution verification is well defined only in terms of specified quantities of interest, which are usually functionals of the full computed solution.

- —Best practice: Clearly define the QOIs for a given VVUQ analysis, including the solution-verification task. Different QOIs will be affected differently by numerical errors.
- —Best practice: Ensure that solution verification encompasses the full range of inputs that will be employed during UQ assessments.
- Principle: The efficiency and effectiveness of code and solution verification can often be enhanced by exploiting the hierarchical composition of codes and mathematical models, with verification performed first on the lowest-level building blocks and then on successively more complex levels.
  - —Best practice: Identify hierarchies in computational and mathematical models and exploit them for code and solution verification. It is often worthwhile to design the code with this approach in mind.
  - —Best practice: Include in the test suite problems that test all levels in the hierarchy.
- The goal of solution verification is to estimate, and control if possible, the error in each QOI *for the problem at hand*.
  - —Best practice: When possible in solution verification, use goal-oriented a posteriori error estimates, which give numerical error estimates for specified QOIs. In the ideal case the fidelity of the simulation is chosen so that the estimated errors are small compared to the uncertainties arising from other sources.
  - —Best practice: If goal-oriented a posteriori error estimates are not available, try to perform self-convergence studies (in which QOIs are computed at different levels of refinement) on the problem at hand, which can provide helpful estimates of numerical error.

Many VVUQ tasks introduce questions that can be posed, and in principle answered, within the realm of mathematics. Validation and prediction introduce additional questions whose answers require judgments from the realm of subject-matter expertise. For validation and prediction, the committee identified several principles and associated best practices, which are detailed in the main text. Some of the more important of these are summarized here:

- Principle: A validation assessment is well defined only in terms of specified quantities of interest (QOIs) and the accuracy needed for the intended use of the model.
  - —Best practice: Early in the validation process, specify the QOIs that will be addressed and the required accuracy.
  - —Best practice: Tailor the level of effort in assessment and estimation of prediction uncertainties to the needs of the application.
- Principle: A validation assessment provides direct information about model accuracy only in the domain of applicability that is "covered" by the physical observations employed in the assessment.
  - —Best practice: When quantifying or bounding model error for a QOI in the problem at hand, systematically assess the relevance of supporting data and validation assessments (which were based on data from different problems, often with different QOIs). Subject-matter expertise should inform this assessment of relevance (as discussed above and in Chapter 7).
  - —Best practice: If possible, use a broad range of sources of physical observations so that the accuracy of a model can be checked under different conditions and at multiple levels of integration.
  - —Best practice: Use "holdout tests" to test validation and prediction methodologies. In such a test some validation data are withheld from the validation process, the prediction machinery is employed to "predict" the withheld QOIs, with quantified uncertainties, and finally the predictions are compared to the withheld data.
  - —Best practice: If the desired QOI was not observed for the physical systems used in the validation process, compare sensitivities of the available physical observations with those of the QOI.
  - —Best practice: Consider multiple metrics for comparing model outputs against physical observations.
- Principle: The efficiency and effectiveness of validation and prediction assessments are often improved by exploiting the hierarchical composition of computational and mathematical models, with assessments beginning on the lowest-level building blocks and proceeding to successively more complex levels.
  - —Best practice: Identify hierarchies in computational and mathematical models, seek measured data that facilitate hierarchical validation assessments, and exploit the hierarchical composition to the extent possible.

—Best practice: If possible, use physical observations, especially at more basic levels of the hierarchy, to constrain uncertainties in model inputs and parameters.
- Principle: Validation and prediction often involve specifying or calibrating model parameters.
  —Best practice: Be explicit about what data/information sources are used to fix or constrain model parameters.
  —Best practice: If possible, use a broad range of observations over carefully chosen conditions to produce more reliable parameter estimates and uncertainties, with less "trade-off" between different model parameters.
- Principle: The uncertainty in the prediction of a physical QOI must be aggregated from uncertainties and errors introduced by many sources, including discrepancies in the mathematical model, numerical and code errors in the computational model, and uncertainties in model inputs and parameters.
  —Best practice: Document assumptions that go into the assessment of uncertainty in the predicted QOI, and also document any omitted factors. Record the justification for each assumption and omission.
  —Best practice: Assess the sensitivity of the predicted QOI and its associated uncertainties to each source of uncertainty as well as to key assumptions and omissions.
  —Best practice: Document key judgments—including those regarding the relevance of validation studies to the problem at hand—and assess the sensitivity of the predicted QOI and its associated uncertainties to reasonable variations in these judgments.
  —Best practice: The methodology used to estimate uncertainty in the prediction of a physical QOI should also be equipped to identify paths for reducing uncertainty.
- Principle: Validation assessments must take into account the uncertainties and errors in physical observations (measured data).
  —Best practice: Identify all important sources of uncertainty/error in validation data—including instrument calibration, uncontrolled variation in initial conditions, variability in measurement setup, and so on—and quantify the impact of each.
  —Best practice: If possible, use replications to help estimate variability and measurement uncertainty.
  —Remark: Assessing measurement uncertainties can be difficult when the "measured" quantity is actually the product of an auxiliary inverse problem—that is, when it is not measured directly but is inferred from other measured quantities.

PROMISING RESEARCH AREAS

After surveying today's VVUQ methods and their mathematical foundations, the committee identified several research topics that offer the promise of improved methods and improved outcomes. The areas identified for verification research are discussed in detail in Chapter 3 and summarized in Chapter 7; they include:

- Development of goal-oriented a posteriori error-estimation methods that can be applied to mathematical models that are more complicated than linear elliptic partial differential equations (PDEs).
- Development of algorithms for goal-oriented error estimates that scale well on massively parallel architectures, especially given complicated grids (including adaptive-mesh grids).
- Development of methods to estimate error bounds when meshes cannot resolve important scales. An example is turbulent fluid flow.
- Development of reference solutions, including "manufactured" solutions, for the kinds of complex mathematical models described above.
- For computational models that are composed of simpler components, including hierarchical models: development of methods that use numerical-error estimates from the simpler components, along with information about how the components are coupled, to produce numerical-error estimates for the overall model.

Research needed to improve uncertainty quantification methodologies is discussed in Chapter 4 and summarized in Chapter 7. Key identified UQ research topics include:

- Development of scalable methods for constructing emulators that reproduce the high-fidelity model results at training points, accurately capture the uncertainty away from training points, and effectively exploit salient features of the response surface.
- Development of phenomena-aware emulators, which would incorporate knowledge about the phenomena being modeled and thereby enable better accuracy away from training points.
- Development of methods for characterizing rare events, for example by identifying input configurations for which the model predicts significant rare events, and estimating their probabilities.
- Development of methods for propagating and aggregating uncertainties and sensitivities across hierarchies of models. (For example, how to aggregate sensitivity analyses across micro-scale, meso-scale, and macro-scale models to give accurate sensitivities for the combined model remains an open problem.)
- Research and development in the compound area of (1) extracting derivatives and other features from large-scale computational models and (2) developing UQ methods that efficiently use this information.
- Development of techniques to address high-dimensional spaces of uncertain inputs.
- Development of algorithms and strategies across the spectrum of UQ-related tasks that can efficiently use modern and future massively parallel computer architectures.

Promising research topics to support validation and prediction are discussed in Chapter 5 and summarized in Chapter 7. Identified topics for validation and prediction include:

- Development of methods and strategies to quantify the effect of subject-matter judgments, which necessarily are involved in validation and prediction, on VVUQ outcomes.
- Development of methods that help to define the "domain of applicability" of a model, including methods that help quantify the notions of near neighbors, interpolative predictions, and extrapolative predictions.
- Development of methods or frameworks that help with the important problem of relating model-to-model differences, among models in an ensemble, to the discrepancy between models and reality.
- Development of methods to assess model discrepancy and other sources of uncertainty in the case of rare events, especially when validation data do not include such events.

Computational modeling and simulation will continue to play key roles in research in engineering and physical sciences (and in many other fields). It already aids scientific discovery, advances understanding of complex physical systems, augments physical experimentation, and informs important decisions. Future advances will be determined in part by how well VVUQ methodology can integrate with the next generation of computational models, high-performance computing infrastructure, and subject-matter expertise. This integration will require that students in these various areas be adequately educated in the mathematical foundations of VVUQ. The committee observes that students in VVUQ-dependent fields are not as well prepared today as they could be to deal with uncertainties that invariably affect problem formulation, software development, and interpretation and presentation of results. As requested by its tasking, the committee identified several actions that could help to address this.

Recommendation: An effective VVUQ education should encourage students to confront and reflect on the ways that knowledge is acquired, used, and updated.

Recommendation: The elements of probabilistic thinking, physical-systems modeling, and numerical methods and computing should become standard parts of the respective core curricula for scientists, engineers, and statisticians.

Recommendation: Researchers should understand both VVUQ methods and computational modeling to more effectively exploit synergies at their interface. Educational programs, including research programs with graduate-education components, should be designed to foster this understanding.

Recommendation: Support for interdisciplinary programs in predictive science, including VVUQ, should be made available for education and training to produce personnel that are highly qualified in VVUQ methods.

Recommendation: Federal agencies should promote the dissemination of VVUQ materials and the offering of informative events for instructors and practitioners.

SUMMARY APPROACH

In summary, the committee has studied VVUQ as it applies to predictive science and engineering, with a focus on the mathematical foundations of VVUQ methodologies. It has identified key principles that it finds helpful and has identified best practices that it has observed in the application of VVUQ to difficult problems in computational science and engineering. It has identified research areas that promise to improve the mathematical foundations that undergird VVUQ processes. Finally, it has discussed changes in the education of professionals and dissemination of information that should enhance the ability of future VVUQ practitioners to improve and properly apply VVUQ methodologies to difficult problems, enhance the ability of VVUQ customers to understand VVUQ results and use them to make informed decisions, and enhance the ability of all VVUQ stakeholders to communicate with each other. The committee offers its observations and recommendations in the hope that they will help the VVUQ community as it continues to improve VVUQ processes and broaden their applications.

1

Introduction

1.1 OVERVIEW AND STUDY CHARTER

Advances in computing hardware and algorithms have dramatically improved the ability to simulate complex processes computationally. Today's simulation capabilities offer the prospect of addressing questions that in the past could be addressed only by resource-intensive experimentation, if at all. However, computational results almost always depend on inputs that are uncertain, rely on approximations that introduce errors, and are based on mathematical models[1] that are imperfect representations of reality. Hence, given some calculated quantity of interest (QOI) from the computational model, the corresponding true physical QOI is uncertain. If this uncertainty—the relationship between the true value of the QOI and the prediction of the computational model—cannot be quantified or bounded, then the computational results have limited value. This report recognizes the ubiquity of uncertainty in computational estimates of reality and the necessity for its quantification. In response to the observation of George Box that "all models are wrong, but some may be useful" (Box and Draper, 1987, p. 424), this report explores how to make models as useful as possible by quantifying how wrong they are.

In a typical computational science and/or engineering analysis, the physical system to be simulated is represented by a mathematical model, which often comprises a set of differential and/or integral equations. The mathematical model is approximated in some way so that the solution of the approximated model can be found by executing a set of algorithms on a computer. For example, derivatives may be approximated by finite differences, series expansions may be truncated, and so on. The computer code's implementation of the algorithms that approximately solve the mathematical model is often called the *computational model* or *computer model*.

As computational science and engineering have matured, the process of quantifying or bounding uncertainties in a computational estimate of a physical QOI has evolved into a small set of interdependent tasks. These tasks are verification, validation, and uncertainty quantification, which are abbreviated as "VVUQ" in this report. Briefly and approximately: verification determines how well the computational model solves the math-model equations, validation determines how well the model represents the true physical system, and uncertainty quantification (UQ) plays important roles in validation and prediction.

---

[1] In this report, a *model* is defined as a representation of some portion of the world in a readily manipulatable form. A *mathematical model* uses the form of mathematical language and equations.

In recognition of the increasing importance of computational simulation and the increasing need to assess uncertainties in computational results, the National Research Council (NRC) was asked to study the mathematical foundations of VVUQ and to recommend steps that will ultimately lead to improved processes. The specific tasking to the Committee on Mathematical Foundations of Verification, Validation, and Uncertainty Quantification is as follows:

- A committee of the National Research Council will examine practices for VVUQ of large-scale computational simulations in several research communities.
- The committee will identify common concepts, terms, approaches, tools, and best practices of VVUQ.
- The committee will identify mathematical sciences research needed to establish a foundation for building a science of verification and validation (V&V) and for improving the practice of VVUQ.
- The committee will recommend educational changes needed in the mathematical sciences community and mathematical sciences education needed by other scientific communities to most effectively use VVUQ.

1.2 VVUQ DEFINITIONS

Figure 1.1 illustrates the different elements of VVUQ and their relationships to the true, physical system, the mathematical model, and the computational model. Uncertainty quantification does not appear explicitly in the figure, but it plays important roles in the processes of validation and prediction.

There is general agreement about the purposes of verification, validation, and uncertainty quantification, but different groups can differ on the details of each term's definition. For purposes of this report the committee adopts the following definitions:

- *Verification*. The process of determining how accurately a computer program ("code") correctly solves the equations of the mathematical model. This includes *code verification* (determining whether the code correctly implements the intended algorithms) and *solution verification* (determining the accuracy with which the algorithms solve the mathematical model's equations for specified QOIs).
- *Validation*. The process of determining the degree to which a model is an accurate representation of the real world from the perspective of the intended uses of the model (taken from AIAA, 1998).
- *Uncertainty quantification* (UQ). The process of quantifying uncertainties associated with model calculations of true, physical QOIs, with the goals of accounting for all sources of uncertainty and quantifying the contributions of specific sources to the overall uncertainty.

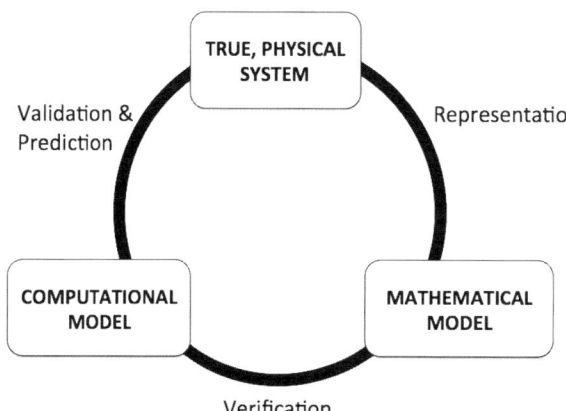

FIGURE 1.1 Verification, validation, and prediction as they relate to the true, physical system, the mathematical model, and the computational model. (Adapted from AIAA [1998].)

INTRODUCTION

In this report, "quantifying uncertainty" in a prediction for a QOI means making a quantitative statement about the values that the QOI for the physical system may take, often in a new, unobserved setting. The statement could take the form of a bounding interval, a confidence interval, or a probability distribution, perhaps accompanied by an assessment of confidence in the statement. Much more is said on this topic throughout this Introduction and the rest of the report.

There is wide but not universal agreement on the terms, concepts, and definitions described above. These and other terms, many of which are potentially confusing terms of art, are discussed in the Glossary (Appendix A).

1.3 SCOPE OF THIS STUDY

1.3.1 Focus on Prediction with Physics/Engineering Models

Mathematical models for the computational simulation of complex real-world processes are a crucial ingredient in virtually every field of science, engineering, medicine, and business. The focus of this report is on physics-based and engineering models, which often provide a strong basis for producing useful extrapolative predictions.

There is a wide range of models, but the science and engineering models on which this report focuses are most commonly composed of integral equations and partial and ordinary differential equations. Each modeling scenario has unique issues and characteristics that strongly affect the implementation of VVUQ. Relevant issues include the following:

- The level of empiricism versus physical laws encoded in the model,
- The availability and relevance of physical data to the predictions required by the scenario,
- The extent of interpolation versus extrapolation needed for the required predictions,
- The complexity of the physical system being modeled, and
- The computational demands of running the computational model.

The modeling framework assumed throughout most of this report is common in science and engineering: a complex physical process or structure is modeled using applied mathematics, typically with a mathematical model consisting of partial differential equations and/or integral equations and with a computational model that solves a numerical approximation of the mathematical model. Referring to the issues listed above, this report considers scenarios in which:

- Models are strongly governed by physical constraints and rules,
- The availability of relevant physical observations is limited,
- Predictions may be required in untested and/or unobserved physical conditions,
- The physical system being modeled may be quite complex, and
- The computational demands of the model may limit the number of simulations that can be carried out.

Of course many of these scenarios are found in other simulation and modeling applications. To this extent, the topics covered in this report are applicable to other modeling applications.

1.3.2 Focus on Mathematical and Quantitative Issues

The focus on mathematical foundations of VVUQ leads this committee to omit important concepts of model evaluation that are more qualitative in nature. The NRC report *Models in Environmental Regulatory Decision Making* (NRC, 2007a) considers a much broader perspective on model evaluation, including topics such as conceptual model formulation, peer review, and transparency that are not considered in this report. *Behavioral Modeling and Simulation: From Individuals to Societies* (NRC, 2007b) considers VVUQ for behavioral, organizational, and societal models.

This report utilizes several examples to illustrate the challenges of executing a mathematically founded VVUQ analysis. Some of these examples have broad implications, and their inclusion in this report is intended as a means to communicate VVUQ concepts and is not intended as a broader discussion of decision making with models. VVUQ activities enhance the decision-making process and are part of a larger group of decision-support tools that include modeling, simulation, and experimentation. The types of decisions discussed in this report can be grouped into two broad categories: (1) decisions that arise as part of the planning and conduct of the VVUQ activities themselves and (2) decisions made with the use of VVUQ results about an application at hand. Chapter 6 discusses the role that VVUQ plays in decision making and includes two examples of how VVUQ fits in with the decision-making process.

1.4 VVUQ PROCESSES AND PRINCIPLES

VVUQ processes must focus on a set of specified QOIs rather than on the full solution of the model. Quantitative statements about errors and uncertainties are well founded only in reference to specified QOIs. For example, there may be much more uncertainty in the estimate of maximum stress than in that of average stress in some component of some structure; thus, no single statement quantifying uncertainty would apply to both of these example QOIs. Similarly, a model may provide accurate estimates for one QOI while yielding inaccurate results for another.

Many physical systems of interest can be viewed as being composed of subsystems, which are themselves composed of sub-subsystems, and so on. Many large-scale computational models are similarly built up from a hierarchy of models, culminating in a complex integrated model. Such hierarchies are illustrated in Figure 1.2.

The advantage of such a hierarchy is that one can begin the VVUQ processes with the lowest-level subsystems, whose models are less complicated and whose data are easier and cheaper to obtain. Once this is done for the lowest-level subsystems and models, the results form a foundation for VVUQ at the next level in the hierarchy, and so on up to the full system and its QOIs.

1.4.1 Verification

Code verification—determining whether the code correctly executes the intended algorithms—presupposes a computer code that has been developed with software-quality engineering practices and results that are appropriate for the intended use. This report assumes that such practices are in play but does not discuss them. Code verification

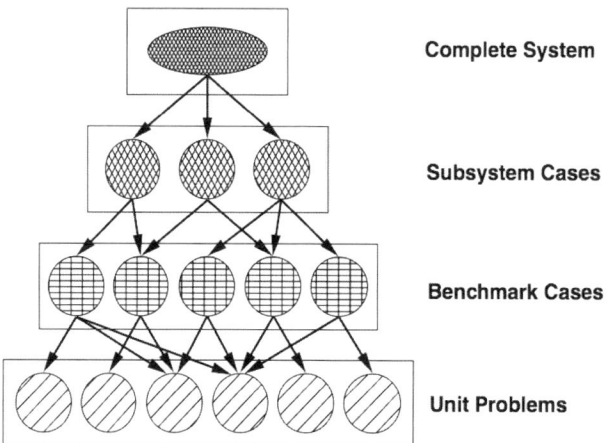

FIGURE 1.2 Validation phases suggested in AIAA (1998), based on hierarchically decomposing a physical system and the models that represent it. Levels in the validation hierarchy range from simple unit problems to benchmark cases to more integrated subsystems and eventually to the complete, integrated system. Verification and uncertainty quantification processes can fruitfully exploit a similar hierarchy.

relies on test problems for which the intended algorithm's correct solution is known. It is challenging to develop a suite of such problems that test all of the relevant algorithms under all relevant combinations of options. Even if a typical complex science or engineering code behaves as expected for a large number of tests, one cannot prove that it has no algorithmic errors.

Solution verification—determining how accurately the numerical algorithm solves the mathematical model equations—must be carried out in terms of specified QOIs, which are usually functionals of the full computed solution. The goal of solution verification is to estimate, and control if possible, the error in each QOI for the simulation problem at hand. The error in a given QOI should not be assumed to be of the same magnitude as the error in another; rather, solution verification must be performed for each QOI individually. Further, the error in a given QOI in a given problem may differ significantly from the error in the same QOI in a slightly different problem.

As noted above, the efficiency and effectiveness of code and solution verification processes can often be enhanced by exploiting the hierarchical composition of codes and solutions, verifying first the lowest-level building blocks and moving successively to more complex levels.

1.4.2 Validation

The process of validation involves comparisons between QOIs computed by the model and corresponding true, physical QOIs inferred from physical observations or experiments. The intended use of the model determines how close the model's QOIs must be to the true QOIs in order for the model to serve its purpose; that is, the intended use determines the requirements on model accuracy. Validation comparisons are conducted relative to these requirements.

In designing experiments for a validation hierarchy, the goal is to carry out a sequence of experiments efficiently, at different levels of the hierarchy, to assess quantitatively the model's ability to predict the QOIs at the full system level. Often in science and engineering applications there are parameters within a model whose values are not known from theory or prior measurements. The best values—or a multivariate probability distribution—for these parameters for a given application must be inferred from new measurements. This inference process is *calibration*. When calibration is required, it is most efficiently and accurately performed at the lowest hierarchical level that provides the needed data. Higher levels introduce confounding factors that make it more difficult to draw quantitative inferences about the parameters in question.

Many factors arise in practice to complicate the validation process. Measurement and inference errors contaminate the QOIs determined from physical observations or experiments. The difference between the computational model and the mathematical model—partly caused by numerical approximations and imperfect iterative convergence—can make it difficult to infer anything about the mathematical model and difficult to determine whether a computational model is getting the "right answer for the right reason." Validation must also take into account the effects of uncertainties in model-input parameters on computed QOIs; thus, the validation process involves UQ processes. These complications are discussed in the Chapter 5, "Model Validation and Prediction."

1.4.3 Prediction

Some experts distinguish model-based predictions at conditions similar to those for which physical observations exist from predictions at new, untested conditions for the physical system (AIAA, 1998; Oberkampf et al., 2004). The terms *interpolative* prediction and *extrapolative* prediction can be used to refer to the former and latter cases, respectively. Suppose that validation assessments have quantified the differences between calculated QOIs and corresponding physical QOIs that were inferred from many previous measurements. Then in the interpolative case, one may be justified in assuming a difference of similar magnitude between the predicted QOI and the true, physical QOI in the untested system. The extrapolative case is more difficult. The new conditions may introduce physical phenomena that are not well modeled, for example, causing the prediction to have greater error than the errors seen in the validation study. That is, it can be risky to assume that the validation study provides a reliable quantitative estimate of the model error for the new problem. If the validation study is assessed *not* to be reliable for the new problem, it is difficult to find a rigorous basis for quantifying the uncertainty in the computationally estimated QOI.

The preceding discussion speaks of interpolative predictions as being relatively straightforward and "safe" and of extrapolative predictions as being relatively difficulty and "risky." While this is intuitively appealing, outside of exceedingly simple settings there is no satisfactory mathematical definition of interpolative or extrapolative categories given the complex science and engineering problems that are addressed in this report. The problems of interest are characterized by large numbers of parameters, which can be viewed mathematically as forming a high-dimensional problem-domain space in which each problem corresponds to a point in the space. Given a finite set of physical observations in such a high-dimensional space, virtually any new problem will be an extrapolation beyond the portion of the domain that is "enclosed" by this set. Even if a new problem can be considered interior to the set of available physical observations, the estimated prediction uncertainty may be unreliable unless the QOI is a smooth function over this domain space. These and related issues are discussed in Chapter 5, "Model Validation and Prediction."

1.4.4 Uncertainty Quantification

The definition adopted in this report for uncertainty quantification describes the overall task of assessing uncertainty in a computational estimate of a physical QOI. This overall task involves many smaller UQ tasks, which are briefly discussed here. This discussion does not explicitly mention exploitation of hierarchical decompositions of the problem, illustrated in Figure 1.2, but such exploitation should be considered when possible in the execution of any VVUQ process.

In the following discussion, it is assumed that preliminaries (code verification, model-parameter calibration if necessary, and validation exercises that have quantified or bounded model error) have been successfully accomplished before UQ begins for a prediction. As mentioned previously in this section, an important question that must be addressed is whether the model error inferred from validation studies is reliably relevant to the new problem being predicted.

The first UQ task is to quantify uncertainties in model inputs, often by specifying ranges or probability distributions. Model inputs include those that do not vary from problem to problem (acceleration of gravity, thermal conductivity of a given material, etc.) as well as those that are problem-dependent (such as boundary and initial conditions).

A key UQ task is to propagate input uncertainties through the calculation to quantify the effects of those uncertainties on the computed QOIs. Whether or not the computational model is an adequate representation of reality, understanding the mapping of inputs to outputs in the model is a key ingredient in the assessment of prediction uncertainty and in gaining an understanding of model behavior. It is conceptually possible to generate a large set of Monte Carlo samples of inputs, run these random input settings through the model, and collect the resulting model outputs to accomplish the forward propagation of input uncertainty. However, the computational demands of the model often preclude the possibility of carrying out a large ensemble of model runs, and the number of points required to sample a high-dimensional input-parameter space densely is prohibitively large. Also, understanding low-probability high-consequence events is difficult using standard Monte Carlo schemes because such events are rarely generated. Current research in mathematical foundations of VVUQ, described in Chapter 4, "Emulation, Reduced-Order Modeling, and Forward Propagation," is focused on approaches for overcoming these challenges for forward propagation of uncertainties.

Another UQ task is quantification of variability in the true physical QOI, which can arise from random processes or from "hidden" variables that are absent from the model. Appropriate methods for quantification depend on the source and nature of the variability.

An important UQ task is the aggregation of uncertainties that arise from different sources. The uncertainties in the QOI that are due to uncertain inputs, true, physical variability, numerical error, and model error must be combined into a quantitative characterization of the overall uncertainty in the computational prediction of a given physical QOI. This UQ challenge and the others mentioned above are discussed in more detail in Chapter 4, "Emulation, Reduced-Order Modeling, and Forward Propagation."

1.4.5 Key VVUQ Principles

A summary of several observations regarding VVUQ follows:

- VVUQ tasks are interrelated. A solution-verification study may incorrectly characterize the accuracy of the code's solution if code verification was inadequate. A validation assessment depends on the assessment of numerical error produced by solution verification and on the propagation of model-input uncertainties to computed QOIs.
- The processes of VVUQ should be applied in the context of an identified set of QOIs. A model may provide an excellent approximation to one QOI in a given problem while providing poor approximations to other QOIs. Thus, the questions that VVUQ must address are not well posed unless the QOIs have been defined.
- Verification and validation are not yes/no questions with yes/no answers but instead are quantitative assessments of differences. Solution verification characterizes the difference between the computational model's solution and the underlying mathematical model's solution. Validation involves quantitative characterization of the difference between computed QOIs and true, physical QOIs.

1.5 UNCERTAINTY AND PROBABILITY

It is unanimously agreed that statistics depends somehow on probability. But, as to what probability is and how it is connected with statistics, there has seldom been such complete disagreement and breakdown of communication since the Tower of Babel. Doubtless, much of the disagreement is merely terminological and would disappear under sufficiently sharp analysis. (Savage, 1954, p. 2)

The hopeful view expressed in the last sentence of the quotation above more than 50 years ago has not been realized. Controversy still abounds concerning the meaning and use of probability and related notions to capture uncertainty (e.g., fuzzy logic). It is not possible to address these issues comprehensively in this report, nor is it the purpose of this study. In the interest of clarity, one framework for reasoning about uncertainty was chosen (out of many possible) and is now described.

The most common method of dealing with uncertainty in VVUQ is through standard probability theory. In this approach unknowns are represented by probability distributions, and rules of probability are used to combine the probability distributions in order to assess the uncertainty in the predictions derived from computer models. This is the approach employed in this report. The approach is capable of synthesizing most, if not all, of the uncertainties that arise in VVUQ, and it offers a straightforward and readily understood means of quantifying uncertainties in model predictions of physical QOIs. This choice is not meant as a judgment that alternative frameworks are invalid or lacking but is meant to provide a single, relatively transparent framework to illuminate VVUQ issues.

Two concerns are often raised with the probabilistic approach. The first concern is the difference between epistemic and aleatoric probability and the combination of these two types of uncertainty. Aleatoric probabilities arise from actual randomness in the real system, and epistemic probabilities typically arise from a lack of knowledge.

- *Example*. Suppose that the manufacturing process for a weapon results in 10 percent nonfunctional weapons; then each weapon randomly has a probability of failure of 0.1 (an aleatoric probability). Alternatively, suppose that the weapon design is based partly on speculative science and the weapons cannot be physically tested. It is judged that there is a 10 percent chance that the science is wrong, in which case none of the weapons would work. Again each particular weapon has a probability of failure of 0.1 (an epistemic probability). But, obviously, the ramifications of these two 0.1 probability statements are very different.

Although a single weapon in this example has a failure probability of 0.1 under either scenario, the joint probability distributions of failure for the weapons are very different. In the aleatoric case, the weapons independently have a 0.1 probability of failure, while in the epistemic case they either all succeed or all fail together, with 0.1 probability of failure. Figure 1.3 shows the joint probability distribution of two weapons in these two different

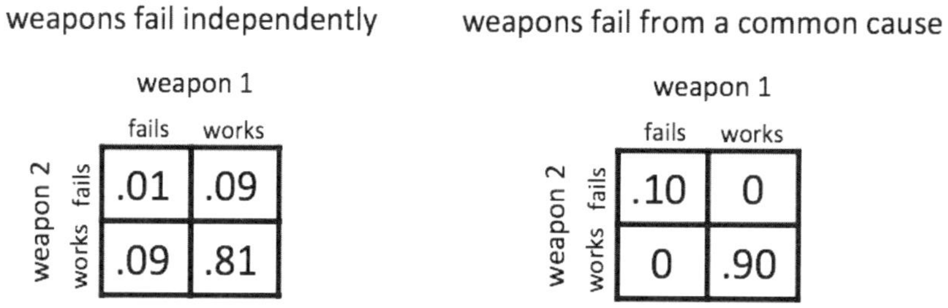

	weapons fail independently	weapons fail from a common cause

Left table (weapons fail independently), weapon 1 / weapon 2:
- weapon 2 fails, weapon 1 fails: .01
- weapon 2 fails, weapon 1 works: .09
- weapon 2 works, weapon 1 fails: .09
- weapon 2 works, weapon 1 works: .81

Right table (weapons fail from a common cause):
- weapon 2 fails, weapon 1 fails: .10
- weapon 2 fails, weapon 1 works: 0
- weapon 2 works, weapon 1 fails: 0
- weapon 2 works, weapon 1 works: .90

FIGURE 1.3 The joint probability distribution for the reliability of two weapons in a stockpile. *Left*: Each weapon works or fails independently, with a 90 percent chance of working. *Right*: A common failure mode gives a 90 percent chance that both weapons work. In this case, either both weapons work or both fail.

situations. The joint distributions communicate the complete picture and clearly differentiate between the two situations. The message here is that standard probability theory does properly distinguish between the two situations, but one must take care in communicating the result. Summarizing a probabilistic description of a prediction with a single metric can be misleading. For example, the chance of failure of a weapon randomly chosen from the stockpile does not address the question of whether or not there is an appreciable chance that the entire stockpile might fail.

Many scientists are reluctant to use the same probability calculus to represent both knowledge uncertainty (epistemic probability) and true randomness (aleatoric probability). This is a centuries-old debate in science and philosophy in which this report will not engage. This report uses the standard probability calculus for both types of uncertainty to illustrate its points, while recognizing that in some applications there may be good reasons to take different approaches.

The second concern often raised with respect to standard probability theory is that it is a "precise" theory, whereas probability judgments are often imprecise. For instance, the precise statement in the above example—that there is a 10 percent chance that the science is wrong—can be questioned; might not a more accurate analysis yield 10.1 percent or 10.01 percent? Here there is less debate philosophically, in that probability judgments are indeed often imprecise. There is no agreement, however, on how to combine imprecise probability judgments into an overall assessment of the uncertainty of predictions. One option is to consider the "worst case" over all the possibilities that are not ruled out. This worst case is often so conservative that it does not provide useful information. However, when worst-case analysis is not overly conservative, it yields a powerful assessment of uncertainty.

Although open to other possibilities, the committee holds the view that currently the use of standard probability modeling is often a reasonable mechanism for producing an overall assessment of accuracy in prediction in VVUQ, and it provides a consistent framework in which this report can illustrate its points. Interval descriptions, such as $0.09 < x < 0.11$, are treated in this report by conversion to standard probability distributions: for example, by treating p as being uniformly distributed over the interval $(0.09, 0.11)$. Other treatments are possible and in some cases may be preferable.

1.6 BALL-DROP CASE STUDY

To help clarify VVUQ concepts, the easily visualized phenomenon of dropping a ball from a tower can be used. Experiments measure the time that it takes the ball to fall from different heights. This simple example, described in Box 1.1 and discussed in this chapter and in Chapter 5, provides a ready physical example for outlining many of the main ideas of VVUQ. This section describes the physical system, posits a simple model for system behavior, and indicates how various sources of uncertainty and bias can affect the model predictions. It is also possible to explore the model in a manner that typifies many applications of physics and engineering and in a way that exposes how uncertainties arise in the resulting prediction for the QOI. These themes recur in many parts of the report.

Box 1.1
Dropping a Bowling Ball from a Tower

The time that it takes for a bowling ball to fall from a tower 100 meters (m) high will be predicted by using experimental drop times obtained from a 60-m tower (Figure 1.1.1(a)). Drop times are recorded for heights of 10, 20, . . . , 50 m, and a validation experiment of dropping a ball from 60 m is also conducted. The uncertainty in the measured drop times is normally distributed about the true time with a standard deviation of 0.1 seconds (s). The quantity of interest (QOI) is the drop time for the bowling ball at a height of 100 m. Since the tower is only 60 m high, a computational model is used to help make this assessment.

The conceptual model incorporates only acceleration due to gravity g, allowing the computational solution used here to be compared to an analytical solution for a verification assessment of bowling ball drop times between 10 m and 100 m.

The physical constant g is assumed to be unknown, but between 8 and 12 m/s² (light lines). The five drop-time measurements (black dots; Figure 1.1.1(b) constrain the uncertainty about g to the probability density given by the dark line.

The drop times produced by the computational model (light lines in Figure 1.1.1(c)) are shown for 11 different values of g. The experimental data are shown with error bars of ±2 standard deviations—the measurement error for the 60 m is not used. The constrained uncertainty for g results in a 95 percent prediction interval for the drop time (as a function of height) depicted by the dark region in Figure 1.1.1(b).

A validation experiment, dropping the bowling ball from 60 m, is conducted so that the prediction can be compared to the measured drop time. The drop-time measurement for 60 m (black dot in Figure 1.1.1(d)) is quite consistent with the prediction and its uncertainty.

A prediction with the uncertainty for the QOI (drop time at 100 m) is given by the light line in Figure 1.1.1(d). The uncertainty accounts for measurement error and parametric uncertainty. Numerical error and model discrepancy are not accounted for in this assessment.

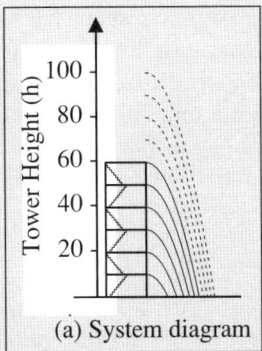
(a) System diagram

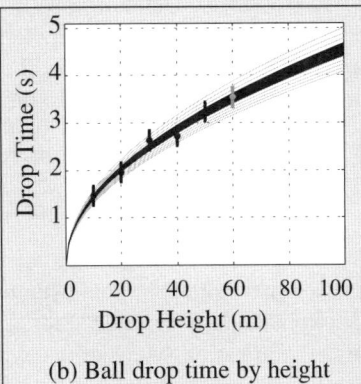
(b) Ball drop time by height

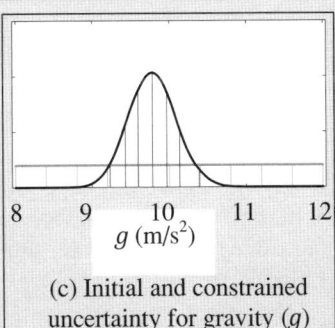
(c) Initial and constrained uncertainty for gravity (g)

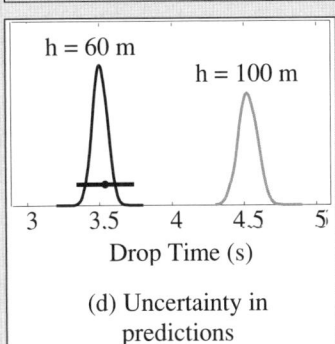
(d) Uncertainty in predictions

FIGURE 1.1.1

1.6.1 The Physical System

With respect to the physical system for the case study, a bowling ball is released from rest at a height h above the ground (see Figure 1.1.1(a) in Box 1.1), and the time that it takes for the ball to reach the ground is recorded. Drop heights of 10 m, 20 m, ..., 60 m are available from the tower. Gravity acts on the ball, causing it to accelerate. The goal is to predict how long it takes for the bowling ball to hit the ground, starting from a height of 100 m. This drop time for a 100-m drop of a bowling ball is the QOI for this experiment. To help address this question, a computational model (described below) and a number of experimentally measured drop times from different heights will be used. But since the tower is only 60 m high, no higher drops are possible.

1.6.2 The Model

The simplest conceivable model for this system assumes that the bowling ball accelerates toward the ground at a constant acceleration given by the gravitational constant g (see Figure 1.1.1(b) in Box 1.1). For now, assume that the exact acceleration due to gravity is unknown—only that this acceleration is between 8 and 12 meters per second per second (m/s^2). Experiments in which the bowling bowl is dropped from different heights will help to reduce this uncertainty. For now, uncertainty about g is described by a uniform prior distribution whose range is from 8 m/s^2 to 12 m/s^2 (see Figure 1.1.1(c), light lines, in Box 1.1).

1.6.3 Verification

The mathematical model is sufficiently simple and the drop time can be computed analytically for any drop height, as long as the value for g is supplied. It is far more common to solve a mathematical system using computational approaches that produce an approximate solution to the system (Morton and Mayers, 2005; Press et al., 2007). Assessing the quality of this approximation is one of the key functions of verification. Much of the work in verification focuses on quantifying or bounding the difference between computationally and mathematically derived solutions. In most applications, this quantification is challenging because the mathematical solution is not available.

1.6.4 Sources of Uncertainty

A number of uncertain quantities that affect the eventual prediction uncertainty for the QOI can be identified. First and most important is the uncertain constant, g—the acceleration due to gravity. Uncertainty regarding this parameter leads to uncertainty in predictions (see Figure 1.1.1(d) in Box 1.1). In other applications, the assessed accuracy of the computational model relative to the mathematical model being solved is an important source of uncertainty (but not here because the mathematical model is relatively straightforward).

The nature, number, and accuracy of the experimental measurements also affect prediction uncertainty. Here the difference between measured and actual drop times should be within 0.2 seconds 95 percent of the time. These deviations, sometimes described as measurement "errors," may be due to the timing process or to variations in the initial position and velocity of the bowling ball as it is released for the drops. The measurement errors are assumed to be independent, directly affecting the updated, or posterior, uncertainty for g. This would not be the case if, for instance, the stopwatch being used ran slightly fast. In that case, all the measured drop times would be systematically low. If it is known that such systematic errors may exist in the measurement process, they can be accounted for probabilistically. When such systematic effects go undetected, obtaining appropriate prediction uncertainties is more difficult.

Inadequacies in the model may also contribute uncertainty to drop-time predictions. This simple model does not account for effects due to air friction. Fortunately, a bowling ball, at the velocities obtained in these experiments, is rather insensitive to the effects of air friction. This would not be the case if a basketball was used instead. Regardless, if very high accuracy is required for the eventual prediction, then additional experiments, more accurate measurements, and/or a more accurate model may be required.

1.6.5 Propagation of Input Uncertainties

One might also like to carry out an uncertainty analysis in which a distribution on the inputs is propagated through the simulation model to give uncertainty about the outputs. This is done in Box 1.1 for both the initial and the constrained distributions for g. Such propagation analyses, which can be carried out in principle using a Monte Carlo simulation, can be very time-consuming when the model is computationally demanding. Dealing with such computational constraints when exploring how the model outputs are affected by input variations is considered in more detail in Chapter 4, "Emulation, Reduced-Order Modeling, and Forward Propagation."

1.6.6 Validation and Prediction

At this point, the experimental observations need to be combined with the computational model in order to obtain more reliable uncertainties regarding the simulation-based prediction for the QOI—the drop time for the bowling ball at 100 m. The drop times can be used to constrain the uncertainty regarding g to give more reliable predictions and uncertainties.

In principle, this inference problem can be tackled using nonlinear regression, from either a likelihood perspective (Seber and Wild, 2003) or a Bayesian perspective (Gelman, 2004). In fact, Figure 1.1.1(c) in Box 1.1 shows the posterior distribution for g resulting from using the experimental measurements to reduce uncertainty and the posterior predictions for the drop times of the bowling ball as a function of height. Here the prediction uncertainty (given by the dark region in Figure 1.1.1(d) in Box 1.1) is due to uncertainty in g. However, this analysis has some drawbacks:

- It requires, at least in principle, many evaluations of the computational model;
- It assumes that the computational model reproduces reality when the appropriate value of g is used; and
- It does not account, in any formal way, for the increased uncertainty that one should expect in predicting the 100-m drop time when only data from heights of 60 m or less are available.

These issues highlight some of the fundamental challenges for the mathematical foundations for VVUQ. Methods for dealing with limited numbers of simulation runs have been a focus of research in VVUQ for the past few decades. However, relatively few approaches for quantifying prediction uncertainty when the computational model has deficiencies have been proposed in the research literature. Predicting the drop time for the bowling ball from 100 m is a good example of how complicated things can get, even for an example as basic as this one.

A helpful experiment—called a validation experiment—tries to assess the model's capability for making a less drastic extrapolation. Here, experiments consisting of drop heights of 10 m, ..., 50 m are used, along with the model, to make a prediction with uncertainty for a drop of 60 m. The prediction and measured results are shown in Figure 1.1.1(d) in Box 1.1, showing strong agreement between the prediction and experimental measurement. Still, the confidence that one should have in the model's prediction for a drop of 100 m remains hard to quantify in any formal manner.

1.6.7 Making Decisions

The ability to model and quantify uncertainties of this system can be used to make decisions about how knowledge of the physical system can be improved. Specifically, what actions will most effectively reduce the uncertainty in predicting the drop time for the bowling ball at 100 m? Actions might include carrying out new experiments, carrying out additional simulations, measuring initial conditions more accurately, improving the experimental timing capabilities, or improving the computational model. For example, the relative merits between extending the tower to 70 m or improving the experimental timing accuracy could be assessed quantitatively, given the available information, costs, and how the new changes are likely to improve uncertainties.

1.7 ORGANIZATION OF THIS REPORT

Following this introductory chapter is the cataloging and discussion of the elements briefly listed in it. Chapter 3 addresses code and solution verification. Chapter 4 addresses the propagation of input uncertainties through the computational model to quantify the resulting uncertainties in calculated QOIs and to carry out sensitivity analyses. Chapter 5 tackles the complex topics of validation and prediction. Chapter 6 addresses the use of computational models and VVUQ to inform important decisions. Chapter 7 discusses today's best practices in VVUQ and identifies research that would improve mathematical foundations of VVUQ. It also discusses VVUQ-related issues in education and offers recommendations for educational changes that would enhance VVUQ capabilities among those who will need to employ them in the future.

1.8 REFERENCES

AIAA (American Institute for Aeronautics and Astronautics). 1998. *Guide for the Verification and Validation of Computational Fluid Dynamics Simulations.* Reston, Va.: AIAA.

Box, G., and N. Draper. 1987. *Empirical Model Building and Response Surfaces.* New York: Wiley.

Gelman, Andrew. 2004. *Bayesian Data Analysis.* Boca Raton, Fla.: CRC Press.

Morton, K.W., and D.F. Mayers. 2005. *Numerical Solution of Partial Differential Equations.* Cambridge, U.K.: Cambridge University Press.

NRC (National Research Council). 2007a. *Models in Environmental Regulatory Decision Making.* Washington, D.C.: The National Academies Press.

NRC. 2007b. *Behavioral Modeling and Simulation: From Individuals to Societies.* Washington, D.C.: The National Academies Press.

Oberkampf, W.L., T.G. Trucano, and C. Hirsch. 2004. Verification, Validation, and Predictive Capability in Computational Engineering and Physics. *Applied Mechanical Reviews* 57:345.

Press, W.H., S.A. Teukolsky, W.T. Vetterling, and B.P. Flannery. 2007. *Numerical Recipes: The Art of Scientific Computing*, Third Edition. New York: Cambridge University Press.

Savage, L.J. 1954. *The Foundations of Statistics.* New York: Wiley.

Seber, G., and C.J. Wild. 2003. *Nonlinear Regression.* Indianapolis, Ind.: Wiley-IEEE.

2

Sources of Uncertainty and Error

2.1 INTRODUCTION

The development of a computational model to predict the behavior of a physical system requires a number of choices from both the analyst, who develops the computational model, and the decision maker, who uses model results to inform decisions.[1] These choices, informed by expert judgment, mathematical and computational considerations, and budget constraints, as well as aspects of the application at hand, all impact how well the computational model represents reality. Each of these choices has the potential to push the computational model away from reality, impacting the validation assessment and contributing to the resulting prediction uncertainty of the model.

In approaching a system whose performance requires a quantitative prediction, the analyst will typically need to make (or at least consider) a series of choices before the analysis can get under way. In particular, thought needs to be given to the following:

- Relevant or interesting measures of the quantity of interest (QOI) from the point of view of any proposed application or decision;
- The underlying quantitative model or theory to use for representing the physical system;
- The adequacy of that underlying model or theory for the proposed application;
- The degree to which the simulation code, as implemented, approximates the results of the underlying model or theory; and
- The fidelity with which the system is represented in the code.

The last four choices in this list are prime sources of analytic uncertainty. In broad terms, the analyst will be uncertain about (1) the choice of theoretical model to use for predicting the QOI, (2) the inherent adequacy of the theoretical model(s) chosen to predict the QOI, and (3) the degree to which any finite computational implementation of a given model for a given problem approximates the actual model solution for that problem. To proceed to a more fine-grained understanding and taxonomy of sources of analytic uncertainty and error, it will be helpful to consider a specific situation—one that is simplified sufficiently to be tractable and yet complex enough to be

---

[1] The terms *analyst* and *decision maker* refer to the roles that the two parties fulfill. Someone who is a decision maker in one context may well be an analyst in another. For example, scientists are often asked to recommend courses of action with regard to funding research projects. In that role, those scientists will be consumers, rather than producers, of simulation-based information.

relevant to the purposes of this study. Tracing the series of analytic choices above in the context of a particular example should help illustrate where potential areas of analytic uncertainty and error can enter into a representative application of simulation-based prediction.

2.2 PROJECTILE-IMPACT EXAMPLE PROBLEM

Consider the situation depicted in Figure 2.1, which is an example adapted from Thompson (1972). The system under consideration consists of a cylindrical aluminum rod impacting the center of a cylindrical aluminum plate at high speed. This kind of system could be informative when trying to understand the behavior of a projectile impacting armor.

At the start of the problem, the cylindrical rod is moving downward at high velocity and is just touching the thick plate (Figure 2.1, left). The image on the right in Figure 2.1 represents a "slice" through the center of the system and uses color to represent density.

Assume that interest centers on the problem of predicting the behavior of the rod and plate system. First, it must be decided which aspects of the system behavior are required to be predicted—the QOIs. There are several possibilities: the depth of penetration of the plate as a function of impactor velocity, the extent of gross damage to the plate, the fine-scale metallographic structure of the plate after impact, the amount of plate material ejected backward after the rod impact, the time-dependent structure of loading and unloading waves during the impact, and so on. In general, the number of possible QOIs that could be considered as reasonable candidates for prediction is large. Which aspects are important is application-dependent—depending, for example, on whether application focus is on improving the performance of the projectile or of the armor plate—and will influence to a large degree the simulation approach taken by the analyst.

A question that emerges at this stage is the degree of accuracy to which the prediction needs to be made. This also depends on the application under consideration. In this rod and plate example, it may be that the primary QOI

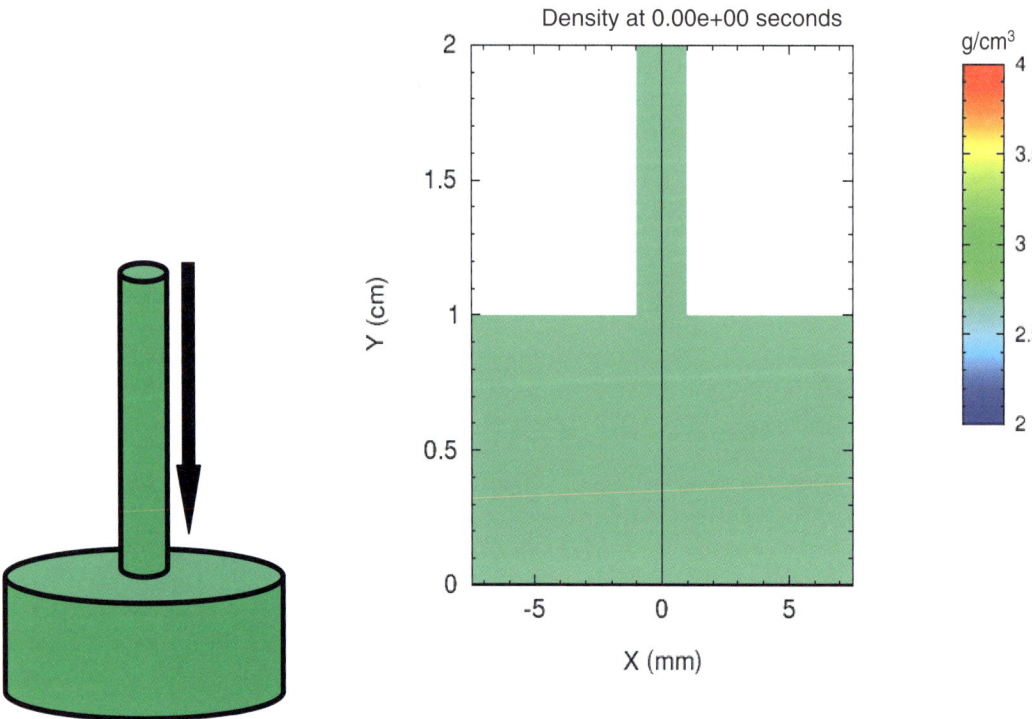

FIGURE 2.1 Aluminum rod impacting a cylindrical aluminum plate.

that is important is the depth of penetration of the rod as a function of impactor velocity, and that the depth only to within a couple of millimeters is the only QOI needed. Having specified the system under consideration, the QOI(s) that require prediction, and the accuracy requirements on that prediction, the analyst can proceed to survey the possible theories or models that are available to estimate the behavior of the system. This aspect of the problem usually requires judgment informed by subject-matter expertise. In this example, the analyst is starting with the equations of solid mechanics. Further, the analyst is assuming that typical impactor velocities are sufficiently large, and the resulting pressures sufficiently high, that the metal rod and plate system can be modeled as a compressible fluid, neglecting considerations of elastic and plastic deformation, material strength, and so on. This is the kind of assumption that draws on relevant background information. Most moderately complex applications rely on such background assumptions (whether or not they are explicitly stated). The specification of the conservation ("governing") equations is given in Box 2.1. The thermodynamic development is not described in any detail, but the problem of specifying thermodynamic properties requires assumptions about model forms—in this example, shown in Box 2.1. The governing equations and other assumptions combine to determine the mathematical model that will be employed, and the specific assumptions made influence the predicted deformation behavior at a fundamental level. The range of validity of any particular set of assumptions is often far more limited than the range of validity of the governing equations. The subsequent predicted results are bounded by the most limiting range of validity.

At this point, strategies for simulating these equations numerically have to be considered—that is, the computational model must be specified. Accomplishing this is far from obvious, and it is greatly complicated by the existence of nonlinear wave solutions (particularly shock waves) to the equations of fluid mechanics. The details of numerical hydrodynamics will not be explored here, but the important point is that the specification of a well-posed mathematical model to represent the physical system is usually just the beginning of any realistic analysis. Strategies for numerically solving the mathematical model on a computer involve significant approximations affecting the computed QOI. The errors resulting from these approximations may be quantified as part of verification activities. Even after a strategy for solving the nonlinear governing equations numerically is chosen, the thermodynamic relations mentioned above have to be computed somehow. This introduces additional approximations as well as uncertain input parameters, and these introduce further uncertainty into the analysis.

Suppose that both a numerical hydrodynamics code (for addressing the governing equations) and the relevant thermodynamic tables (for addressing the thermodynamic assumptions) are readily available. The next issue to consider is the resolution of the spatial grid to use for discretizing the problem. Finite grid resolution introduces numerical error into the computed solution, which is yet another source of error that contributes to the prediction uncertainty for the QOIs. The uncertainty in the QOIs due to numerical error is studied and quantified in the solution verification phase of the verification, validation, and uncertainty quantification (VVUQ) process. After having made all these choices, the simulation can (finally) be run. The computed result is shown in Figure 2.2.

The code predicts that the rod penetrates to a depth of about 0.7 cm in the plate at the time shown in the simulation. The result also shows strong shock compression of the plate that depends in complicated ways on the location within the plate. Finally, there is evidence of interesting fine-scale structure (due to complex interactions of loading and unloading waves) on the surface of the plate.

Any, or all, of these results may be of interest to both the analyst and the decision maker. The exact solution of the mathematical model cannot be obtained for this problem because the propagation of nonlinear waves through real materials in realistic geometries is generally not solvable analytically. Had the analyst been unable to run the code, he or she might still have been able to give an estimate of the QOI, but that estimate would, most likely, have been vastly more uncertain—and less quantifiable—than an estimate based on a reasonably accurate simulation. A primary goal of a VVUQ effort is to estimate the prediction uncertainty for the QOI, given that some computational tools are available and some experimental measurements of related systems are also available. The experimental measurements permit an assessment of the difference between the computational model and reality, at least under the conditions of the available experiments, a topic that is discussed throughout Chapter 5, "Model Validation and Prediction." Note that uncertainties in experimental measurements also impact this validation assessment. An important point to realize, for the purposes of this discussion, is that the computational model results all depend on the many choices made in developing the computational model, each potentially pushing the computed QOI away from its counterpart from the true, physical system. Different choices at any of the stages

Box 2.1
Equations for Conservation of Mass, Momentum, and Energy

To be explicit about the form that the rod and plate analysis might take, it is useful to write the governing equations of continuum mechanics. These are laws of conservation of mass, momentum, and energy:[1]

$$\frac{d\rho}{dt} = -\rho \nabla_k u_k$$

$$\rho \frac{du_k}{dt} = \nabla_l \tau_{kl}$$

$$\rho \frac{dE}{dt} = \tau_{kl} \dot{\varepsilon}_{kl}$$

In these equations, ρ denotes the mass density, E the internal energy per unit volume, u_k is the velocity vector written in Cartesian coordinates, and τ_{kl} is the stress tensor, again written in Cartesian coordinates. ∇_k denotes the differentiation with respect to the kth spatial direction.

The strain-rate tensor in the energy equation is given by:

$$\dot{\varepsilon}_{kl} = \frac{1}{2}(\nabla_k u_l + \nabla_l u_k)$$

and the stress tensor in the momentum and energy equations by:

$$\tau_{kl} = P\delta_{kl} + \tau_{kl}^{diss}$$

These equations are quite general and do not depend on assumptions about elastic or plastic deformation, material strength, and the like. They are simply conservation laws, and their introduction is accompanied by very little uncertainty, at least for this application.

However, the specific form taken by the dissipative component of the stress tensor, τ^{diss}, relies on approximations, particularly involving the thermodynamic specification of the system. If, following the analysis mentioned above, it is decided that the rod and plate system will be modeled as a viscous fluid, then the viscous component of the stress tensor can be expressed as a function of the shear and bulk viscosities, η_s and η_v:

$$\tau_{kl}^{diss} = \tau_{kl}^{NS} = 2\eta_s \dot{\varepsilon}_{kl} + (2\eta_v - \frac{2}{3}\eta_s) \dot{\varepsilon}_{rr} \delta_{kl}$$

These viscosities (and the internal energy) must be expressed in terms of independent thermodynamic variables, which require approximations that introduce still more uncertainty into the predicted quantities of interest.

---
[1] For these equations and notable conventions, see Wallace (1982).

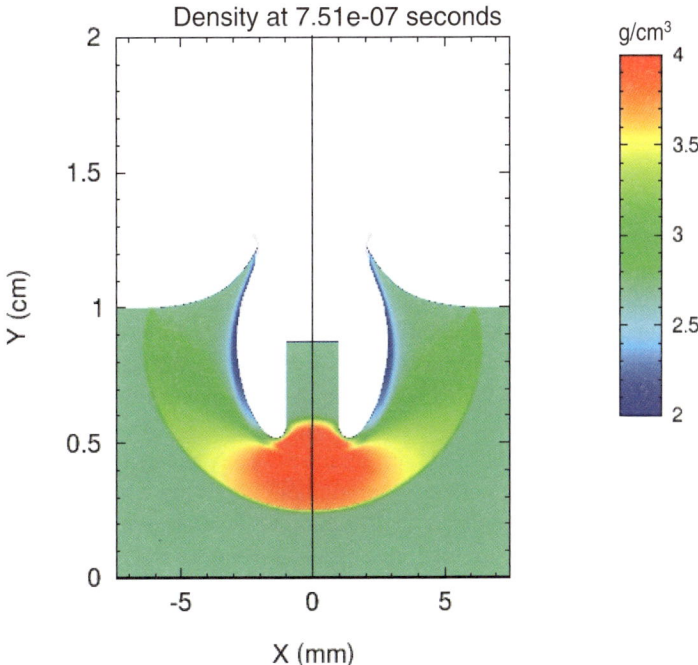

FIGURE 2.2 Aluminum rod at the end of simulation. The figure shows a slice through the system (which began as two cylinders). Color represents density.

discussed above would produce different results from those pictured in Figure 2.2. The results, however, are not equally sensitive to every choice—some choices have greater influence than others on the computational results. A major goal of any uncertainty quantification (UQ) effort is to disentangle these sensitivities for the problem at hand. The problem discussed here exhibits many of the sources of uncertainty that are likely to be present in most computational analyses of physical or engineered systems. It is worthwhile to extract some of the salient themes from this cursory overview.

2.3 INITIAL CONDITIONS

One uncertainty that the discussion of this example problem did not emphasize is uncertainty in the initial data for the problem. It was simply assumed that the initial dimensions, velocities, densities, and so forth for the combined rod and plate system were known arbitrarily well. Many problems that are commonly simulated do not enjoy this luxury. For example, hydrodynamic simulations of incompressible turbulent flows often suffer from large uncertainties in the initial conditions of the physical system. In principle, these uncertainties must be parameterized in some way and then processed through the simulation code so that different parameter settings describe different initial conditions. If the rest of the modeling process is perfect—something that the above discussion indicates is far from likely—uncertainty in the initial conditions will be the dominant contributor to prediction uncertainty in the QOI. In this case, methods mentioned in Chapter 4, "Emulation, Reduced-Order Modeling, and Forward Propagation," for the propagation of input uncertainty would describe the prediction uncertainty. The main thrust of the preceding discussion of the example problem has been to point out the many possible additional sources of uncertainty over and above uncertainty in the initial conditions.

Finding: Common practice in uncertainty quantification has been to focus on the propagation of input uncertainties through the computational model in order to quantify the resulting uncertainties in output quantities of interest, with substantially less attention given to other sources of uncertainty.

There is a need for improved theory, methods, and tools for identifying and quantifying uncertainties from sources other than uncertain computational inputs and for aggregating the uncertainties that arise from all sources.

2.4 LEVEL OF FIDELITY

Fidelity of the numerical representation of a complex system is another aspect of many simulation-based predictions that is not well exemplified by the preceding example problem. The geometry of a cylindrical rod impacting a plate is not a particularly difficult geometry to capture in modern computer codes. Many other problems of interest, however, have significant difficulties in capturing all the geometric intricacies of the real system. In many cases, rather complex three-dimensional geometries are simplified to two-, one-, or even zero-dimensional numerical representations.[2] But even if the full dimensionality of the problem is retained, it is often the case—especially for intricate systems like a car engine—that many of the pieces of a complex system have to be either ignored or greatly simplified in the code just to get the problem generated or for the code to run stably in a reasonable amount of time.

These decisions are present in most simulation analyses. They are typically made on the basis of expert judgment. One method of determining the effect of a judgment regarding what to neglect or simplify is to try to include a more complete representation and see if it affects the answer significantly. However, if every aspect of the real world could be accurately represented, that is generally what the analyst would do; so in some cases, more detailed modeling is not feasible. An alternative method is to simplify the representation sufficiently that it can be accommodated by the code and computer that are available and accept the impact that the approximate representation has on the simulation output. This is often the only reasonably available strategy for getting an answer and so is frequently the one taken. However, this method also leaves a great deal of latitude to the analyst in making choices on how to represent the system numerically and on judging whether or not the simplifications compromise the simulation results or provide information of utility to the decision maker.

2.5 NUMERICAL ACCURACY

As mentioned briefly in the discussion of the example problem, every analyst confronts the issues of uncertainty in the numerical solution of the equations embodied in the code. Three different broad, but potentially overlapping, categories of uncertainty can be distinguished:

- Inadequate algorithms,
- Inadequate resolution, and
- Code bugs.

Algorithmic inadequacies and resolution inadequacies stem from a common cause: most mathematical models in computational science and engineering are formulated using continuous variables and so have the cardinality of the real numbers; but all computers are finite. In most cases, derivatives are finite differences and integrals are finite sums. Different algorithms for approximating, say, a differential equation on a computer will have different convergence properties as the spatial and temporal resolution are increased. Because the spatial and temporal resolutions or number of Monte Carlo samples of a probability distribution are generally fixed by the computational hardware available, certain algorithms will usually be more appropriate than others. There may be additional factors that also favor one algorithm over another. It is a time-honored warning, though, that every algorithm, no matter how well it performs generally, may have an Achilles' heel—some weakness that will cause it to stumble when faced with a certain problem. Unfortunately, these weaknesses are usually found in one of two ways: first, by comparing the code against a problem with a known solution; second, by comparing the code against reality. In either case, checks are made to see where the code is found wanting. The problem with checking against a known solution—while undoubtedly a useful and important procedure—is that it is very difficult to determine whether

---

[2] An example of a much-used zero-dimensional code is the ORIGEN2 nuclear reactor isotope depletion code. See Croff (1980).

or not the weakness thus revealed in the simple problem will play an important role in the true application. Again, one should note that any judgments made by the analyst on how to factor in algorithmic issues are typically both complex and somewhat subjective.

Resolution inadequacies are slightly different in that they can be checked, at least in principle, by increasing (or decreasing) the spatial and temporal resolutions to estimate the impact of resolution on the simulation result. Notice, again, that here the analyst will be restricted by computational expense and, perhaps, by intrinsic algorithmic limitations.

Finally, code bugs are the rule, not the exception, in all codes of any complexity. Software quality engineering and code verification are finely developed fields in their own right. Nightly regression suites (software to test whether software changes have introduced new errors), manufactured solutions,[3] and extensive testing are all attempts to ensure that as many of the lines of code as possible are error-free. It may be trite but is certainly true to say that a single bug may be sufficient to ensure that the best computational model in the world, run on the most capable computing platform, produces results that are utterly worthless.

Finding: The VVUQ process would be enhanced if the methods and tools used for VVUQ and the methods and algorithms in the computer model were designed to work together. To facilitate this, it is important that code developers and model developers learn the basics of VVUQ methodologies and that VVUQ method developers learn the basics of computational science and engineering. The fundamentals of VVUQ, including strengths, weaknesses, and underlying assumptions, are important components in the education of analysts who are responsible for making predictions with quantified uncertainties.

2.6 MULTISCALE PHENOMENA

Systems that require computational simulation to predict their evolution in time and space are most frequently intrinsically nonlinear. Fluid dynamics is a paradigmatic example. The hallmark of a nonlinear system is that the dynamics of the system couple many different degrees of freedom. This phenomenon is seen in the example problem of a rod impacting a plate. Steady nonlinear waves, such as shocks, critically depend on the nonlinearity of the dynamical equations. A shock couples the large-scale motions of the solid or fluid system to the small-scale regions where the work done by the viscous stresses is dissipated. Multiscale phenomena always complicate a simulation. As was seen very briefly above, the net effect of the nonlinearity is to present the modeler with a choice of options:

- *Option 1*: Directly model all the scales of interest, or
- *Option 2*: Choose a cutoff scale—that is, a scale below which the phenomena will not be represented directly in the simulation, replacing the physical model with another model that is cutoff-dependent.

Each alternative has associated advantages and disadvantages. Option 1 is initially appealing, but it usually dramatically increases the computational expense of even a simple problem, and the time spent in setting up such a simulation might be unwarranted for the task at hand. Moreover, it often introduces additional parameters, governing system behavior at different scales, that need to be calibrated beforehand, frequently in regimes where those parameters are poorly known. The uncertainties induced by these uncertain parameters might exceed the additional fidelity that one could expect from a more complete model. Option 2, on the other hand, reduces the computational expense by limiting the degrees of freedom that the simulation resolves, but it usually involves the construction of physically artificial numerical models whose form is, to some extent, unconstrained and which usually have their own adjustable parameters. In practice, such effective models are almost always tuned, or calibrated, to reproduce the behavior of selected problems for which the correct large-scale behavior is known. The

---

[3] *Manufactured solutions* refers to the process of postulating a solution in the form of a function, followed by substituting it into the operator characterizing the mathematical model in order to obtain (domain and boundary) source terms for the model equations. This process then provides an exact solution for a model that is driven by these source terms (Knupp and Salari, 2003; Oden, 1994).

example problem presents the analyst with this choice. In keeping with the usual practice in compressible-fluid numerical hydrodynamics, Option 2 was followed. The dynamics were modeled with an effective numerical model, an artificial viscosity. Each of these options presents its own type of uncertainty.

2.7 PARAMETRIC SETTINGS

If it is assumed that the form of the small-scale physics is known (Option 1), then the main residual uncertainty of Option 1 is *parametric*, that is, due to uncertainty about the correct values to assign to the parameters in the correct physics model. The fortunate aspect of parametric uncertainty is that many methods (e.g., Bayesian, maximum-likelihood) of parameter calibration are available for estimating parameters from experimental data, if such data are available. Again, however, one has to compare the computational expense involved in the direct simulation of many different length scales with the expense of obtaining additional data.

2.8 CHOOSING A MODEL FORM

Option 2, choosing a cutoff scale, introduces a potentially important contributor to model discrepancy: *model-form uncertainty*. When an effective model is created to mimic the physics of the length scales that are deleted from the simulation, the form of the model—the types of terms that enter into the equations—is usually not fully determined by the requirement that the effective model reproduce the physics of a selected subset of large-scale motions. For example, the Von Neumann-Richtmyer artificial viscosity method (Von Neumann and Richtmyer, 1950) (and its descendants) is a way of introducing some effects of viscosity into simulations based on equations that do not otherwise represent the root causes of viscosity. These methods are a good illustration of potential model-form error in that they can accurately propagate planar shocks in a single material—the physics that they are designed to replicate—while failing (in different ways) to model higher-dimensional, or multimaterial, hydrodynamic situations.

A typical strategy to use in constructing effective models is to require that the effective model obey whatever symmetries happen to be present in the full model. In fluid dynamics, for example, one would prefer to have a subgrid model possess the symmetries of the full Navier-Stokes equations.[4] Often, however, retaining the full symmetry group is not practicable. Even when it is, however, the symmetry group still usually permits an infinite number of possible terms that may satisfy the symmetry requirements. Which of these terms need to be retained is, unfortunately, often problem-dependent, a fact that can create difficulties for general-purpose codes. Moreover, methods for expressing model-form error, and assessing its impact on prediction uncertainty, are in their infancy compared to methods for addressing parametric uncertainty. Sometimes, however, it is possible to parameterize a model-form uncertainty, allowing it to be treated as a parametric uncertainty. This type of reduction can occur by using parameters to control the appearance of the terms in an effective model after the imposition of symmetry requirements.

2.9 SUMMARY

The simplified, but still representative, simulation problem discussed here is presented in the hope of identifying at least some sources of error and uncertainty in model-based predictions. It should be clear that the impact of including these different effects is to, on the whole, push a computed QOI away from its counterpart in the physical system, increasing the resulting prediction uncertainty. However, the manner in which this increase occurs will depend considerably on the details of the simulation and uncertainty models under consideration. Some uncertainties will add in an independent manner; others may have strong positive or negative correlations. It is important to realize that some uncertainties may be limited by the availability of relevant experimental data. The most straightforward example occurs when one has data that constrain the possible values of one or more input parameters to the simulation code being used in the analysis. Clearly, the details of such interactions and constraints and the

---

[4]These symmetries are discussed in Frisch (1995, Chapter 2).

structure of the analysis are closely interrelated. The sources of uncertainty and error identified here, while almost certainly not an exhaustive list, will likely be present in most simulation-based analyses. Developing quantitative methods to address such a wide variety of uncertainty and error also presents difficult challenges, and exciting research opportunities, to the verification and validation (V&V) and UQ communities.

2.10 CLIMATE-MODELING CASE STUDY

The previous discussion noted that uncertainty is pervasive in models of real-world phenomena, and climate models are no exception. In this case study, the committee is not judging the validity or results of any of the existing climate models, nor is it minimizing the successes of climate modeling. The intent is only to discuss how VVUQ methods in these models can be used to improve the reliability of the predictions that they yield and provide a much more complete picture of this crucial scientific arena.

Climate change is at the forefront of much scientific study and debate today, and UQ should be central to the discussion. To set the stage, one of the early efforts at UQ analysis for a climate model is described (from Stainforth et al., 2005). This study considered two types of input uncertainties for climate models: uncertainty in the initial conditions for the climate model (i.e., the posited initial state of climate on Earth, which is very imprecisely known, particularly with respect to the state of the ocean) and uncertainty in the climate model parameters (i.e., uncertainty in coefficients of equations defining the climate model, related to unknown or incompletely represented physics). It is standard in weather forecasting, and relatively common in climate modeling, to deal with uncertainty in the initial conditions by starting with an *ensemble* (or set) of possible initial states and propagating this ensemble through the model. It is less common to attempt to deal with the uncertainty in the model parameters, although this has been considered in Forest et al. (2002) and Murphy et al. (2004), for example. The climate model studied was a version of a general circulation model from the United Kingdom Met Office[5] consisting of the atmospheric model HadAM3[6] coupled to a mixed-layer ocean. Out of the many parameters in the model, six parameters—relating to the way clouds and precipitation are represented in the model—were varied over three plausible values (chosen by scientific experts). For emphasis, note that the standard output that one would see from runs of this climate model would be from the run that uses the central value of each of the three possible parameters. Figure 2.3(b) (from Stainforth et al., 2005) indicates the effect of this parameter uncertainty in the prediction of the effect of CO_2-doubling on global mean temperature change over a 15-year period. This discussion does not consider the calibration and control phases of the analyses, but note the considerable uncertainty in the final prediction of global mean temperature change at the end of the final CO_2-doubling phase. If the climate model were simply run at its usual setting for the parameters, one would obtain only one result, corresponding to an increase of 3.4 degrees. Note also that the initial state of the climate is unknown; to reflect this, a total of 2,578 runs of the climate model were made, varying both the model parameters and the initial conditions. The uncertainty in the final prediction of global mean temperature is then indicated by the larger final spread in Figure 2.3(a).

This discussion only scratches the surface of UQ analysis for climate change. The variation allowed in this study in the model parameters was modest, and only six of the many model parameters were varied. Uncertainty caused by model resolution and incorrect or incomplete structural physics (e.g., the form of the equations) also needs to be considered. The former could be partially addressed by studying models at differing resolutions, and the latter might be approached by comparing different climate models (see, e.g., Smith et al., 2009); it is quite likely, however, that differing climate models make many of the same modeling approximations, and, because of the finite resolution inherent in today's computers, can only imperfectly resolve difficult topological features. The extent of the effect of chaotic behavior is also poorly understood in climate change (the dust bowl of the 1930s may well have been a chaotic event that would not appear in climate models—see Seager et al., 2009), regional effects are likely to be even more variable, and uncertainty as to future significant changes in forcing parameters (e.g., the actual level of CO_2 increase) should also be taken into account. Of course, future CO_2 levels depend on human action, which adds more difficulties to an already complex problem.

---

[5] See www.metoffice.gov.uk. Accessed August 19, 2011.
[6] See Pope et al. (2000).

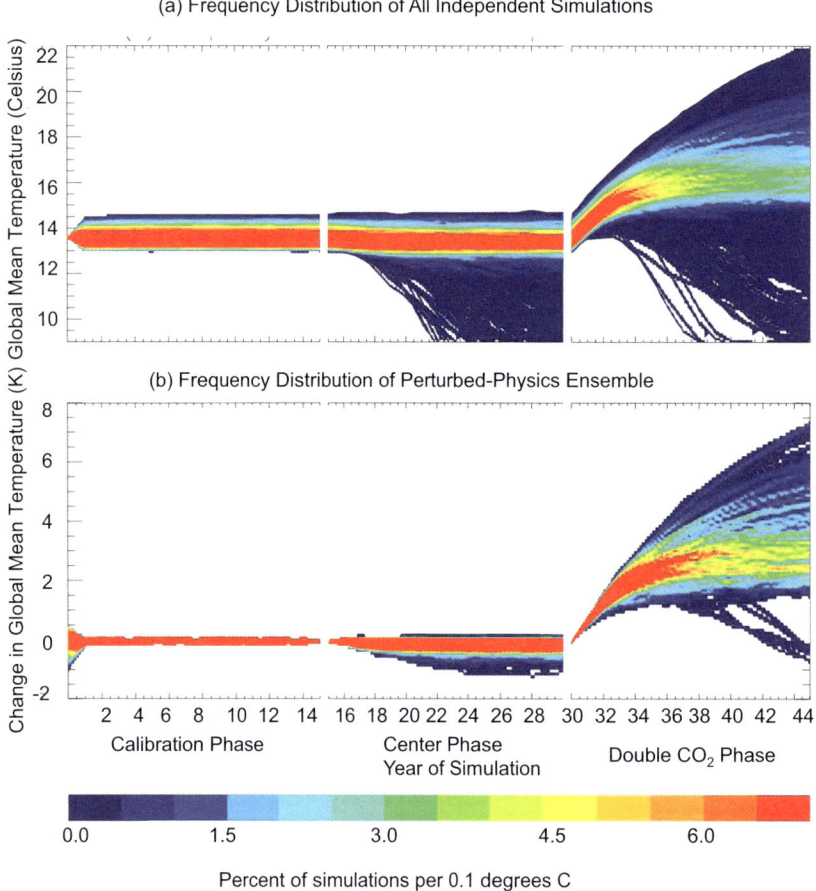

FIGURE 2.3 After calibration and control phases, the effect on global mean temperature of 15 years of doubling of CO_2 forcing is considered (a) when both initial conditions and model parameters are varied and (b) when only model parameters are varied. SOURCE: Stainforth et al. (2005).

2.10.1 Is Formal UQ Possible for Truly Complex Models?

The preceding case study provides a useful venue for elaboration on a more general issue concerning UQ that was raised by committee discussions with James McWilliams (University of California, Los Angeles), Leonard Smith (London School of Economics), and Michael Stein (University of Chicago). The general issue is whether formal validation of models of complex systems is actually feasible. This issue is both philosophical and practical and is discussed in greater depth in, for example, McWilliams (2007), Oreskes et al. (1994), and Stainforth et al. (2007). As discussed in this report, carrying out the validation process is feasible for complex systems. It depends on clear definitions of the intended use of the model, the domain of applicability, and specific QOIs. This is discussed further in Chapter 5, "Model Validation and Prediction."

Several factors make uncertainty quantification for climate models difficult. These include:

1. If the system is hugely complex, the model is, by necessity, only a rough approximation to reality. That this is the case for climate models is indicated by the difficulty of simultaneously tuning climate models to fit numerous outputs; global climate predictions can often be tuned to match various statistics from

nature, but then will be skewed for others. Formally quantifying the uncertainty caused by all of the simplifying assumptions needed to construct a climate model is daunting and involves the participation of people—not just the use of software—because one has to vary the model structure in meaningful ways. One must understand which simplifying assumptions that were made in model construction were rather arbitrary and have plausible alternatives.

Another fundamental challenge is that the resulting variations in model predictions do not naturally arise probabilistically, so that it is unclear how to describe the uncertainty formally and to combine it with the other uncertainties in the problem. For the latter, it might be practically necessary to use a probabilistic representation, but an understanding of the limitations or biases in such an approach is needed.

2. Simultaneously combining all of the sources of uncertainty relating to climate models in order to assess the uncertainty of model predictions is highly challenging, both at the formulation level and in technical implementation.[7]
3. There is a need to make decisions regarding climate change before a complete UQ analysis will be available. This, of course, is not unique to climate modeling, but is a feature of other problems like stewardship of the nuclear stockpile. This does not mean that UQ can be ignored but rather that decisions need to be made in the face of only partial knowledge of the uncertainties involved. The "science" of these kinds of decisions is still evolving, and the various versions of decision analysis are certainly relevant.

2.10.2 Future Directions for Research and Teaching Involving UQ for Climate Models

In spite of the challenges in the formal implementation of UQ in climate modeling, the committee agrees that understanding uncertainties and trying to assess their impact is a crucial undertaking. Some future directions for research and teaching that the committee views as highly promising are the following:

1. It is important to instill an appreciation that modeling truly complex systems is a lengthy process that cannot proceed without exploring mathematical and computational alternatives. Instead, it is a process of learning the behavior of the system being modeled and understanding limitations in the ability to predict such systems.
2. One must recognize that it is often possible to perform formally UQ only on part of a complex system and that one should develop ways in which this partial UQ can be used. For instance, one might state bounds on the predictive uncertainty arising from the partial UQ and then list the other sources of uncertainty that have not been analyzed. And, as always, the domain of applicability of the UQ analysis must be stated—for example, it may be that the assessment is only valid for predictions on a continental scale and for a horizon of 20 years. This recognition is even more important when one realizes that even a complex system such as a climate model is ultimately itself just a component of a more complex system, involving paleoclimate models, space-time hierarchical models, and so on.
3. There is a tendency among modelers to always use the most complex model available, but such a model can be too expensive to run to allow for UQ analysis. It would be beneficial to instill an appreciation that use of a smaller model, together with UQ analysis, is often superior to a single analysis with the most complex model. In weather forecasting, for instance, it was discovered that running a smaller model with ensembles of initial conditions gave better forecasts than those from a single run of a bigger model. This discovery has carried over to some extent in climate modeling—climate models are kept at a level of complexity wherein an ensemble of initial conditions can be considered, but allowance for UQ with other uncertainties is also needed.

---

[7] See, for example, R. Knutti, R. Furer, C. Tebaldi, J. Cermak, and G.A. Mehl. 2010. Challenges in Combining Projections in Multiple Climate Models. *Journal of Climate* 23(10):2739-2758.

2.11 REFERENCES

Croff, A.G. 1980. *ORIGEN2—A Revised and Updated Version of the Oak Ridge Isotope Generation and Depletion Code*. Oak Ridge National Laboratory Report ORNL-5621. Oak Ridge, Tenn.: Oak Ridge National Laboratory.

Forest, C.E., P.H. Stone, A.P. Sokolov, M.R. Allen, and M.D. Webster. 2002. Quantifying Uncertainties in Climate System Properties with the Use of Recent Climate Observations. *Science* 295(5552):113-117.

Frisch, U. 1995. *Turbulence*. Cambridge, U.K.: Cambridge University Press.

Knupp, P., and K. Salari. 2003. *Verification of Computer Codes in Computational Science and Engineering*. Boca Raton, Fla.: Chapman and Hall/CRC.

McWilliams, J.C. 2007. Irreducible Imprecision in Atmospheric and Oceanic Simulations. *Proceedings of the National Academy of Sciences* 104:8709-8713.

Murphy, J.M., D.M.H. Sexton, D.N. Barnett, G.S. Jones, M.J. Webb, M. Collins, and D.A. Stainforth. 2004. Quantification of Modelling Uncertainties in a Large Ensemble of Climate Change Simulations. *Nature* 430(7001)768-772.

Oden, J.T. 1994. Error Estimation and Control in Computational Fluid Dynamics. Pp. 1-23 in *The Mathematics of Finite Elements and Applications*. J.R. Whiteman (Ed.). New York: Wiley.

Oreskes N., K. Shrader-Frechette, and K. Belitz. 1994. Verification, Validation, and Confirmation of Numerical Models in the Earth Sciences. *Science* 263:641-646.

Pope, V.D., M.L. Gallani, P.R. Rowntree, and R.A. Stratton. 2000. The Impact of New Physical Parameterizations in the Hadley Centre Climate Model: HadAM3. *Climate Dynamics* 16(2-3):123-146.

Seager, R., Y. Kushnir, M.F. Ting, M. Cane, N. Naik, and J. Miller. 2009. Would Advance Knowledge of 1930s SSTs Have Allowed Prediction of the Dust Bowl Drought? *Journal of Climate* 22:193-199.

Smith, R., C. Tebaldi, D. Nychka, and L. Mearns. 2009. Bayesian Modeling of Uncertainty in Ensembles of Climate Models. *Journal of the American Statistical Association* 104:97-116.

Stainforth, D.A., T. Aina, C. Christensen, M. Collins, N. Faull, D.J. Fram, J.A. Kettleborough, S. Knight, A. Martin, J.M. Murphy, C. Piani, D. Sexton, L.A. Smith, R.A. Spicer, A.J. Thorpe, and M.R. Allen. 2005. Uncertainty in Predictions of the Climate Response to Rising Levels of Greenhouse Gases. *Nature* 433:403-406.

Stainforth, D.A., M.R. Allen, E. Tredger, and L.A. Smith. 2007. Confidence, Uncertainty and Decision-Support Relevance in Climate Predictions. *Philosophical Transactions of the Royal Society A: Mathematical, Physical and Engineering Sciences* 365(1857):2145-2161.

Thompson, P. 1972. *Compressible-Fluid Dynamics*. New York: McGraw-Hill.

Von Neumann, J., and R.D. Richtmyer. 1950. A Method for the Numerical Calculation of Hydrodynamic Shocks. *Journal of Applied Physics* 21(3):232-237.

Wallace, D.C. 1982. Theory of the Shock Process in Dense Fluids. *Physical Review A* 25(6):3290-3301.

3

Verification

3.1 INTRODUCTION

In Chapter 1, verification is defined as the process of determining how accurately a computer program ("code") correctly solves the equations of a mathematical model. This includes code verification (determining whether the code correctly implements the intended algorithms) and solution verification (determining the accuracy with which the algorithms solve the mathematical model's equations for specified quantities of interest [QOIs]).

In this chapter verification is discussed in detail. The chapter begins with overarching remarks, then discusses code verification and solution verification, and closes with a summary of verification principles.

Many large-scale computational models are built up from a hierarchy of models, as is illustrated in Figure 1.1. Opportunities exist to perform verification studies that reflect the hierarchy or collection of these models into the integrated simulation tool. Both code and solution verification studies may benefit by taking advantage of this composition of submodels because the submodels may be more amenable to a broader set of techniques. For example, code verification can fruitfully employ "unit tests" that assess whether the fundamental software building blocks of a given code correctly execute their intended algorithms. This makes it easier to test the next level in the code hierarchy, which relies on the previously tested fundamental units. As another example, solution verification in a calculation that involves interacting physical phenomena is aided and enhanced if it is performed for the individual phenomena. In the following sections, it is implicitly assumed that this principle of hierarchical decomposition is to be followed when possible.

The processes of verification and validation (V&V) and uncertainty quantification (UQ) presuppose a computational model or computer code that has been developed with software quality engineering practices appropriate for its intended use. Software quality assurance (SQA) procedures provide an essential foundation for the verification of complex computer codes and solutions. The practical implementation of SQA in software development may be approached using risk-based grading with respect to software quality. The basic notion of risk-based grading is straightforward—the higher the risk associated with the usage of the software, the greater the care that must be taken in the software development. This approach attempts to balance the programmatic drivers, scientific and technological creativity, and quality requirements. Requirements governing the development of software may manifest themselves in regulations, orders, guidance, and contracts; for example, the Department of Energy (DOE) provides documents detailing software quality requirements in (DOE, 2005). Standards are available that outline activities that are elements for ensuring appropriate software quality (American National Standards Institute, 2005). The discipline of software quality engineering presents a breadth of practices that can be put into place.

For example, the DOE provides a suggested set of goals, principles, and guidelines for software quality practices (DOE, 2000). The use of these practices can be tailored to the development environment and application area. For example, pervasive use of software configuration management and regression testing is the state of practice in many scientific communities.

3.2 CODE VERIFICATION

Code verification is the process of determining whether a computer program ("code") correctly implements the intended algorithms. Various tools for code verification and techniques that employ them have been proposed (Roache, 1998, 2002; Knupp and Salari, 2003; Babuska, 2004). The application of these processes is becoming more prevalent in many scientific communities. For example, the computer model employed in the electromagnetics case study described in Section 4.5 uses carefully verified simulation techniques. Tools for code verification include, but are not limited to, comparisons against analytic and semianalytic (i.e., independent, error-controlled) solutions and the "method of manufactured solutions." The latter refers to the process of postulating a solution in the form of a function, followed by substituting it into the operator characterizing the mathematical model in order to obtain (domain and boundary) source terms for the model equations. This process then provides an exact solution for a model that is driven by these source terms (Knupp and Salari, 2003; Oden, 2003).

The comparison of code results against independent, error-controlled ("reference") solutions allows researchers to assess, for example, the degree to which code implementation achieves the expected solution to the mathematical system of equations. Because the reference solution is exact and the code implements numerical approximation to the exact solution, one can test convergence rates against those predicted by theory. As a separate, complementary activity one can often construct a reference solution to the *discretized* problem, providing an independent solution of the computational model (*not* the mathematical model). This verification activity allows assessment of correctness in the pre-asymptotic regime. To make such reference solutions mathematically tractable, typically simplified model problems (e.g., ones with lower dimensionality and with simplified physics and geometry) are chosen. However, there are few analytical and semianalytical solutions for more complex problems. There is a need to develop independent, error-controlled solutions for increasingly complex systems of equations—for example, those that represent coupled-physics, multiscale, nonlinear systems. Developing such solutions that are relevant to a given application area is particularly challenging. Similarly, developing manufactured solutions becomes more challenging as the mathematical models become more complex, since the number of terms in the source expressions grows in size and complexity, requiring great care in managing and implementing the source terms into the model.

Another challenge is the need to construct manufactured solutions that expose different features of the model relevant to the simulation of physical systems, for example, different boundary conditions, geometries, phenomena, nonlinearities, or couplings. The method of manufactured solutions is employed in the verification methodology used in the Center for Predictive Engineering and Computational Sciences (PECOS) study described in Section 5.9. Finally as indicated in Section 5.9, manufactured or analytical solutions should reproduce known challenging features of the solution, such as boundary layers, effects of interfaces, anisotropy, singularities, and loss of regularity.

Some communities employ cross-code comparisons (in which different codes solve the same discretized system of partial differential equations [PDEs]) and refer to this practice as verification. Although this activity provides valuable information under certain conditions and can be useful to ensure accuracy and correctness, this activity is not "verification" as the term is used in this report. Often the reference codes being compared are not themselves verifiable. One of the significant challenges in cross-code comparisons is that of ensuring that the codes are modeling identical problems; these codes tend to vary in the effects that they include and in the way that the effects are included. It may be difficult to simulate identical physics processes and problems. One needs to model the same problem for the different codes; ideally the reference solution is arrived at using a distinct error-controlled numerical technique.

Upon completion of a code-verification study, a statement can be made about the correctness of the implementation of the intended algorithms in the computer program under the conditions imposed by the study (e.g., selected QOIs, initial and boundary conditions, geometry, and other inputs). To ensure that the code continues to be subjected to verification tests as changes are made to it, verification problems from the study are typically incorporated in a code-development test suite, established as part of the software quality practices.

In practice, regression test suites are composed of a variety of tests. Since the problems used for regression suites may be constructed for that purpose, it may be that a solution in continuous variables is known or that the problem is constructed with a particular solution in mind (manufactured solution). Such a suite may include verification tests (including comparison against continuous solutions, solutions to a discretized version of the problem, and manufactured solutions) in addition to other types of tests—unit tests (tests of a particular part or "unit" of the code), integration tests (tests of integrated units of the code), and user-acceptance tests. Performing these verification studies and augmenting test suites with them help to ensure the quality and pedigree of the computer code. A natural question arises as to the sufficiency and adequacy of regularly passing these test suites as code development continues. Various metrics (coverage metrics) have been developed to measure the ability of the tests to cover various aspects of the code, including source lines of computer code, functions of the code, and features of the code (Jones, 2000; Westfall, 2010). Care must be taken when interpreting the results of these coverage metrics, particularly for scientific code development, because coverage metrics tend to measure the execution of a particular portion of the computer code, independent of the input values. Uncertainty quantification studies explore a broad variety of input parameters, which may result in unexpected results from algorithms and physics models, even those that have undergone extensive testing.

Some software development teams have found utility in employing static analysis tools, including those that incorporate logic-checking algorithms, for source code checking (Ayewah et al., 2008). Static code analysis is a software analysis technique that is performed without actually executing the software. Modern static analysis tools parse the code in a way similar to what compilers do, creating a syntax tree and database of the entire program's code, which is then analyzed against a set of rules or models (Cousot, 2007). On the basis of those rules, the analysis tool can create a report of suspected defects in the code. The formalism associated with these rules allows potential defects to be categorized according to severity and type. Since the analysis tools have access to a database of the entire source code, defects that are a combination of source code statements in disparate locations in the code implementations can be identified (e.g., allocation of memory in one portion of the implementation without release of that memory prior to returning control flow). Such tools can aid in verifying that the source code implementation is a correct realization of the intended algorithms. However, to date these tools have been able to answer only limited questions about codes of limited complexity. The expansion of these tools to science and engineering simulation software, and the kinds of questions that they may ultimately be able to answer, remain topics for further study.

3.3 SOLUTION VERIFICATION

Solution verification is the process of determining or estimating the accuracy with which algorithms solve the mathematical-model equations for the given QOIs. A breadth of tools have been developed for solution verification; these include, but are not limited to, a priori and a posteriori error estimation[1] and grid adaptation to minimize numerical error. The most sophisticated solution-verification techniques incorporate error estimation with error control (by means of h-, p-, or r-adaptivity) in the physics simulation.

QOIs are typically expressed as functionals of the fully computed solution across the problem domain. The solution of the mathematical-model equations is often a set of dependent-variable values that are evaluated at a large number of points in a space defined by a set of independent variables. For example, the dependent variables could be pressure, temperature, and velocity; the independent variables could be position and time. In the usual case, it is not the value at each point that is of interest but rather the more aggregated quantities—such as the average pressure in a space-time region—that are functionals of the complete solution.

---

[1] A priori error estimation is done by an examination of the model and computer code; a posteriori error estimation is done by an examination of the results of the execution of the code.

Finding: Solution verification (determining the accuracy with which the numerical methods in a code solve the model equations) is useful only in the context of specified quantities of interest, which are usually functionals of the fully computed solution.

The accuracy of the computed solution may be very different for different pointwise quantities and for different functionals. It is important to identify the QOIs because the discretization and resolution requirements for predicting these quantities may vary (e.g., predicting an integral quantity across the spatial domain may be less restrictive than predicting local, high-order derivatives). The PECOS case study presented in Section 5.9 employs quantities of interest as a fundamental aspect of solution verification.

Solution verification is a matter of numerical-error estimation, the goal being to estimate the error present in the computational QOI relative to an exact QOI from the underlying mathematical model. While code verification considers generic formulations of simplified problems within a class that the code was designed to treat, solution verification pertains to the specific, large-scale modeling problem that is at the center of the simulation effort, with specific inputs (boundary and initial conditions, constitutive parameters, solution domains, source terms) and outputs (the QOIs). The goal of the solution-verification process is to estimate and control the error *for the simulation problem at hand*. The most sophisticated realization of this technique is online during the solution process to ensure that the actual delivered numerical solution from the code is a reliable estimate of the true solution to the underlying mathematical model. Not all discretization techniques and simulation problems lend themselves to this level of sophistication. Solution verification may also employ relevant reference solutions, self-convergence, and other techniques for estimating and controlling numerical error prior to performing the simulation at hand.

Solution-verification practices may employ independent, error-controlled ("reference") solutions. Comparing code results against reference solutions allows researchers to estimate, for example, the numerical error introduced by the discretized equations being employed and to assess the order of accuracy. Maintaining second-order, or even first-order, convergence can be challenging in complex, nonlinear, multiphysics simulations. Obtaining reference solutions that are demonstrably relevant to the simulation at hand is challenging, particularly for highly complex large-scale models; thus, the application of this approach to solution verification is limited. More reference solutions that exhibit features of the phenomena of interest are needed for complex problems, including those with strong nonlinearities, coupled physical phenomena, coupling across scales, and stochastic behavior. Generating relevant reference solutions for these and other complex, nonlinear, multiphysics problems would extend the breadth of problems for which this approach to solution verification could be employed.

Solution verification may also be performed by using the code itself to produce high-resolution reference solutions—a practice referred to as performing "self-convergence" studies. If rigorous error estimates are available, they can be used to extrapolate successive discrete calculations to estimate the infinite-resolution solution. In the absence of such an error estimate, the highest-resolution simulation may be used as the reference "converged" solution. Such studies can be used to assess the rate at which self-convergence is achieved in the QOIs and to inform the discretization and resolution requirements to control numerical error for the simulation problem at hand. This approach has the benefit that the complexity of the problem of interest is limited only by the capabilities of the code being studied and the computer being used, removing the limitation typically imposed by requiring independent error-controlled solutions.

Methods of numerical-error estimation generally fall into two categories: a priori estimates and a posteriori estimates. The former, when available, can provide useful information on the convergence rates obtainable as approximation parameters (e.g., mesh sizes) are refined, but they are of little use in quantifying numerical error in quantities of interest. A posteriori estimates aim to achieve quantitative estimates of numerical error (Babuska and Stromboulis, 2001; Ainsworth and Oden, 2000). Methods in this category include explicit and implicit residual-based methods for global error measures, variants of Richardson extrapolation,[2] superconvergence recovery methods, and goal-oriented methods based on adjoint solutions. The recent development of goal-oriented adjoint-based methods, in particular, has produced methods that are capable of yielding, in many cases, guaranteed bounds on

---

[2]*Richardson extrapolation* is a numerical technique used to accelerate the rate of convergence of a sequence. See Brezinski and Redivo-Zaglia (1991).

errors for specific applications and quantities of interest (Becker and Rannacher, 2001; Oden and Prudhomme, 2001; Ainsworth and Oden, 2000). This research incorporates ingredients needed to control numerical errors for QOIs in simulation problems at hand. Recent extensions include the ability to treat error in stochastic PDEs (Almeida and Oden, 2010) and errors for multiscale and multiphysics problems (Estep et al., 2008; Oden et al., 2006), including molecular and atomistic models and combined atomistic-to-continuum hybrids (Bauman et al., 2009). Parallel adaptive mesh refinement (AMR) methods (Burstedde et al., 2011) have been integrated with adjoint-based error estimators to bring error estimation to very large-scale problems on parallel supercomputers (Burstedde et al., 2009).

Despite the recent successes in the development of goal-oriented, adjoint-based methods, a number of challenges remain. These include the development of two-sided bounds for a broader class of problems (beyond elliptic PDEs), further extensions to stochastic PDEs, and the generalizations of adjoints for nonsmooth and chaotic systems. Additionally, challenges remain for the development of the theory and scalable implementations for error estimation on adaptive and complex meshes (e.g., p- and r-adaptive discretizations and AMR). The development of rigorous a posteriori error estimates and adaptive control of all components of error for complex, multiphysics, multiscale models is an area that will remain ripe for research in computational mathematics over the coming decade.

Finding: Methods exist for estimating tight two-sided bounds for numerical error in the solution of linear elliptic PDEs. Methods are lacking for similarly tight bounds on numerical error in more complicated problems, including those with nonlinearities, coupled physical phenomena, coupling across scales, and stochasticity (as in stochastic PDEs).

The results of the solution-verification process help in quantitatively estimating the numerical error impacting the quantity of interest. More sophisticated techniques allow the numerical error to be controlled in the simulation, allowing researchers to target a particular maximum tolerable error and adapt the simulation to meet that requirement, provided sufficient computational time and memory are available. Typically, such adaptation controls the discretization error present in the model. Such techniques can then lead to managing the total error in a simulation, including discretization error as well as errors introduced by iterative algorithms and other approximation techniques.

Finding: Methods exist for estimating and controlling spatial and temporal discretization errors in many classes of PDEs. There is a need to integrate the management of these errors with techniques for controlling errors due to incomplete convergence of iterative methods, both linear and nonlinear. Although work has been done on balancing discretization and solution errors in an optimal way in the context of linear problems (e.g., McCormick, 1989; Rüde, 1993), research is needed on extending such ideas to complex nonlinear and multiphysics problems.

Managing the total error of a solution offers opportunities to gain efficiencies throughout the verification, validation, and uncertainty quantification (VVUQ) process. As with other aspects of the verification process, managing total error is best done in the context of the use of the model and the QOIs. The error may be managed differently for a "best physics" estimate to a particular quantity of interest versus an ensemble of models being used to train a reduced-order model. Managing the total error appropriately throughout the VVUQ study may allow improvement of the turnaround time of the study.

3.4 SUMMARY OF VERIFICATION PRINCIPLES

Important principles that emerge from the above discussion of code and solution verification are as follows:

- Solution verification must be done in terms of specified QOIs, which are usually functionals of the full computed solution.

- The goal of solution verification is to estimate and control, if possible, the error in each QOI *for the simulation problem at hand*.
- The efficiency and effectiveness of code- and solution-verification processes can often be enhanced by exploiting the hierarchical composition of codes and solutions, verifying first the lowest-level building blocks and then moving successively to more complex levels.
- Verification is most effective when performed on software developed under appropriate software quality practices. These include software-configuration management and regression testing.

3.5 REFERENCES

Ainsworth, M., and J.T. Oden. 2000. *A Posteriori Error Estimation in Finite Element Analysis*. New York: Wiley Interscience.

Almeida, J., and T. Oden. 2010. Solution Verification, Goal-Oriented Adaptive Methods for Stochastic Advection-Diffusion Problems. *Computer Methods in Applied Mechanics and Engineering* 199(37-40):2472-2486.

American National Standards Institute. 2005. *American National Standards: Quality Management Systems—Fundamentals and Vocabulary* ANSI/ISO/ASQ Q9001-2005. Milwaukee, Wisc.: American Society for Quality.

Ayewah, N., D. Hovemeyer, J.D. Morgenthaler, J. Penix, and W. Pugh. 2008. Using Static Analysis to Find Bugs. *IEEE Software* 25(5):22-29.

Babuska, I. 2004. Verification and Validation in Computational Engineering and Science: Basic Concepts. *Computer Methods in Applied Mechanics and Engineering* 193:4057-4066.

Babuska, I., and T. Strouboulis. 2001. *The Finite Element Method and Its Reliability*. Oxford, U.K.: Oxford University Press.

Bauman, P.T., J.T. Oden, and S. Prudhomme. 2009. Adaptive Multiscale Modeling of Polymeric Materials with Arlequin Coupling and Goals Algorithms. *Computer Methods in Applied Mechanics and Engineering* 198:799-818.

Becker, R., and R. Rannacher. 2001. An Optimal Control Approach to a Posteriori Error Estimation in Finite Element Methods. *Acta Numerica* 10:1-102.

Brezinski, C., and M. Redivo-Zaglia. 1991. *Extrapolation Methods*. Amsterdam, Netherlands: North-Holland.

Burstedde, C., O. Ghattas, T. Tu, G. Stadler, and L. Wilcox. 2009. Parallel Scalable Adjoint-Based Adaptive Solution of Variable-Viscosity Stokes Flow Problems. *Computer Methods in Applied Mechanics and Engineering* 198:1691-1700.

Burstedde, C., L.C. Wilcox, and O. Ghattas. 2011. Scalable Algorithms for Parallel Adaptive Mesh Refinement on Forests of Octrees. *SIAM Journal on Scientific Computing* 33(3):1103-1133.

Cousot, P. 2007. The Role of Abstract Interpretation in Formal Methods. Pp. 135-137 in *SEFM 2007, 5th IEEE International Conference on Software Engineering and Formal Methods,* London, U.K., September 10-14. Mike Hinchey and Tiziana Margaria (Eds.). Piscataway, N.J.: IEEE Press.

DOE (Department of Energy). 2000. *ASCI Software Quality Engineering: Goals, Principles, and Guidelines*. DOE/DP/ASC-SQE-2000-FDRFT-VERS2. Washington, D.C.: Department of Energy.

DOE. 2005. *Quality Assurance*. DOE O414. Washington, D.C.: Department of Energy.

Estep, D., V. Carey, V. Ginting, S. Tavener, and T. Wildey. 2008. A Posteriori Error Analysis of Multiscale Operator Decomposition Methods for Multiphysics Models. *Journal of Physics: Conference Series* 125:1-16.

Jones, C. 2000. *Software Assessments, Benchmarks, and Best Practices*. Upper Saddle River, N.J.: Addison Wesley Longman.

Knupp, P., and K. Salari. 2003. *Verification of Computer Codes in Computational Science and Engineering*. Boca Raton, Fla.: Chapman and Hall/CRC.

McCormick, S. 1989. *Multilevel Adaptive Methods for Partial Differential Equations*. Philadelphia, Pa.: Society for Industrial and Applied Mathematics.

Oden, J.T. 2003. Error Estimation and Control in Computational Fluid Dynamics. Pp. 1-23 in *The Mathematics of Finite Elements and Applications*. J.R. Whiteman (Ed.). New York: Wiley.

Oden, J.T., and A. Prudhomme. 2001. Goal-Oriented Error Estimation and Adaptivity for the Finite Element Method. *Computer Methods in Applied Mechanics and Engineering* 41:735-756.

Oden, J.T., S. Prudhomme, A. Romkes, and P. Bauman. 2006. Multi-Scale Modeling of Physical Phenomena: Adaptive Control of Models. *SIAM Journal on Scientific Computing* 28(6):2359-2389.

Roache, P. 1998. *Verification and Validation in Computational Science and Engineering*. Albuquerque, N.Mex.: Hermosa Publishers.

Roache, P. 2002. Code Verification by the Method of Manufactured Solutions. *Journal of Fluids Engineering* 124(1):4-10.

Rüde, U. 1993. *Mathematical and Computational Techniques for Multilevel Adaptive Methods*. Philadelphia, Pa.: Society for Industrial and Applied Mathematics.

Westfall, L. 2010. *Test Coverage: The Certified Software Quality Handbook*. Milwaukee, Wisc.: ASQ Quality Press.

4

Emulation, Reduced-Order Modeling, and Forward Propagation

Computational models simulate a wide variety of detailed physical processes, such as turbulent fluid flow, subsurface hydrology and contaminant transport, hydrodynamics, and also multiphysics, as found in applications such as nuclear reactor analysis and climate modeling, to name only a few examples. The frequently high computational cost of running such models makes their assessment and exploration challenging. Indeed, continually exercising the simulator to carry out tasks such as sensitivity analysis, uncertainty analysis, and parameter estimation is often infeasible. The analyst is instead left to achieve his or her goals with only a limited number of calls to the computational model or with the use of a different model altogether. In this chapter, methods for computer model emulation and sensitivity analysis are discussed.

Two types of emulation settings, designed to solve different, but related, problems, must be distinguished. The first type attempts to approximate the dependence of the computer model outputs on the inputs. In this case, the uncertainty comes from not having observed the full range of model outputs or from the fact that another model is used in place of the costly computational model of interest. These emulators include regression models, Gaussian process (GP) interpolators, and Lagrangian interpolations of the model output, as well as reduced-order models.

The second type of emulation problem—discussed in Section 4.2—is similar, with the additional considerations that the input parameters are now themselves uncertain. So, the aim is to emulate the distribution of outputs, or a feature thereof, under a prespecified distribution of inputs. Statistical sampling of various types (e.g., Monte Carlo sampling) can be an effective tool for mapping uncertainty in input parameters into uncertainty in output parameters (McKay et al., 1979). In its most fundamental form, sampling does not retain the functional dependence of output on input, but rather produces quantities that have been averaged simultaneously over all input parameters. Alternatively, approaches such as polynomial chaos attempt to leverage mathematical structure to achieve more efficient estimates of quantities of interest (QOIs). Indeed, polynomial chaos expansions can lead to a more tractable representation of the uncertainty of the QOIs, which can then be explored using mathematical or computational means.

Finally, a few details should be noted before proceeding. The methods discussed—such as emulation, reduced-order modeling, and polynomial chaos expansions—use output produced from ensembles of simulations carried out at different input settings to capture the behavior of the computational model, the aim being to maximize the amount of information available for the uncertainty quantification (UQ) study given a limited computational budget. The term *emulator* is most often used to describe the first type of emulation problem and is the terminology adopted hereafter. Furthermore, unless otherwise indicated, the computational models are assumed to be deterministic. That is, code runs repeated at the same input settings will yield the same outputs.

4.1 APPROXIMATING THE COMPUTATIONAL MODEL

Representing the input/output relationships in a model with a statistical surrogate (or emulator) and using a reduced-order model are two broad methods effectively used to reduce the computational cost of model exploration. For instance, a reduced-order model (Section 4.1.2) or an emulator (Section 4.1.1) can be used to stand in place of the computer model when a sensitivity analysis is being conducted or uncertainty is propagating across the computer model (see Section 4.2 and the example on electromagnetic interference phenomena in Section 4.5). Of course, as with any approximation, there is a reduction in the accuracy of the estimates obtained, and the trade-off between accuracy and cost needs to be considered by the analyst.

4.1.1 Computer Model Emulation

In settings in which the simulation model is computationally expensive, an emulator can be used in its place. The computer model is generally viewed as a black box, and constructing the emulator can be thought of as a type of response-surface modeling exercise (e.g., Box and Draper, 2007). That is, the aim is to establish an approximation to the input-output map of the model using a limited number of calls of the simulator.

Many possible parametric and nonparametric regression techniques can provide good approximations to the computer-model response surface. For example, there are those that interpolate between model runs such as GP models (Sacks et al., 1989; Gramacy and Lee, 2008) or Lagrange interpolants (e.g., see Lin et al., 2010). Approaches that do not interpolate the simulations, but which have been used to stand in place of the computer models, include polynomial regression (Box and Draper, 2007), multivariate adaptive regression splines (Jin et al., 2000), projection pursuit (see Ben-Ari and Steinberg, 2007, for a comparison with several methods), radial basis functions (Floater and Iske, 1996), support vector machines (Clarke et al., 2003), and neural networks (Hayken, 1998), to name only a few. When the simulator has a stochastic or noisy response (Iooss and Ribatet, 2007), the situation is similar to the sampling of noisy physical systems in which random error is included in the statistical model, though the variability is likely to also depend on the inputs. In this case, any of the above models can be specified so that the randomness in the simulator response is accounted for in the emulation of the simulator.

Some care must be taken when emulating deterministic computer models if one is interested in representing the uncertainty (e.g., a standard deviation or a prediction interval) in predictions at unsampled inputs. To deal with the difference from the usual noisy settings, Sacks et al. (1989) proposed modeling the response from a computer code as a realization of a GP, thereby providing a basis for UQ (e.g., prediction interval estimation) that most other methods (e.g., polynomial regression) fail to do. A correlated stochastic process model with probability distribution more general than that of the GP could also be used for this interpolation task. A significant benefit of the Gaussian model is the persistence of the tractable Gaussian form following conditioning of the process at the sampled points and the representation of uncertainty at unsampled inputs.

Consider, for example, the behavior of the prediction intervals in Figure 4.1. Figure 4.1(a) shows a GP fit to deterministic computer-model output, and Figure 4.1(b) shows the same data fit using ordinary least squares regression with the set of Legendre polynomials. Both representations emulate the computer model output fairly well, but the GP has some obvious advantages. Notice that the fitted GP model passes through the observed points, thereby perfectly representing the deterministic computational model at the sampled inputs. In addition, the prediction uncertainty disappears entirely at sites for which simulation runs have been conducted (the prediction is the simulated response). Furthermore, the resulting prediction intervals reflect the uncertainty one would expect from a deterministic computer model—zero predictive uncertainty at the observed input points, small predictive uncertainty close to these points, and larger uncertainty farther away from the observed input points.

In spite of the aforementioned advantages, GP and related models do have shortcomings. For example, they are challenging to implement for large ensemble sizes. Many response-surface methods (e.g., polynomial regression or multivariate adaptive regression splines) can handle much larger sample sizes than the GP can and are computationally faster. Accordingly, adapting these approaches so that they can have the same sort of inferential advantages, as shown in Figure 4.1, as those of the GP in the deterministic setting is a topic of ongoing and future research.

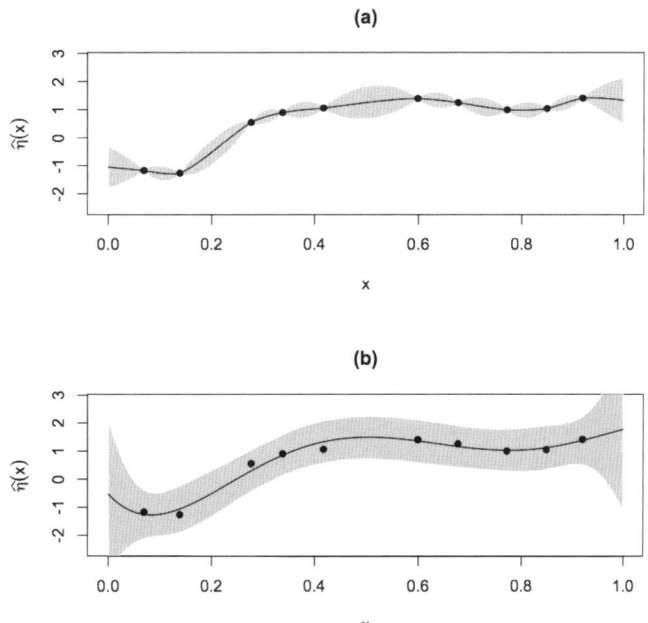

FIGURE 4.1 Two prediction intervals fit to a deterministic computer model: (a) Gaussian process and (b) ordinary least squares regression with Legendre polynomials.

In the coming years, as computing resources become faster and more available, emulation will have to make use of very large ensembles over ever-larger input spaces. Existing emulation approaches tend to break down if the ensemble size is too large. To accommodate larger and larger ensemble sizes, new computational schemes that are suitable for high-performance computing architectures will be required for fitting emulators to computer-model output and producing predictions from these emulators.

Finding: Scalable methods do not exist for constructing emulators that reproduce the high-fidelity model results at each of the N training points, accurately capture the uncertainty away from the training points, and effectively exploit salient features of the response surface.

Finally, most of the current technology for fitting response surfaces treats the computational model as a black box, ignoring features such as continuity or monotonicity that might be present in the physical system being modeled. Augmented emulators that incorporate this phenomenology could provide better accuracy away from training points (Morris, 1991). Current approaches that make use of derivative or adjoint information are examples of emulators that include additional information about the phenomena being modeled.

Finding: Many emulators are constructed only with knowledge about values at training points but do not otherwise include knowledge about the phenomena being modeled. Augmented emulators that incorporate this phenomenology could provide better accuracy away from training points.

4.1.2 Reduced-Order Models

An alternative to emulating the computational model is to use a reduced-order version of the forward model—which is itself a reduced-order model of reality. There are several approaches to achieve this, with projection-based model-reduction techniques being the most developed. These techniques aim to identify within the state space a low-dimensional subspace in which the "dominant" dynamics of the system reside (i.e., those dynamics important for accurate representation of input-output behavior). Projecting the system-governing equations onto this low-dimensional subspace yields a reduced-order model. With appropriate formulation of the model reduction problem, the basis and other elements of the projection process can be precomputed in an off-line phase, leading to a reduced-order model that is rapid to evaluate and solve for new parameter values.

The substantial advances in model reduction over the past decade have taken place largely in the context of forward simulation and control applications; however, model reduction has a large potential to accelerate UQ applications. The challenge is to derive a reduced model that (1) yields accurate estimates of the relevant statistics, meaning that in some cases the model may need to well represent the entire parameter space of interest for the UQ task at hand, and (2) is computationally efficient to construct and solve.

Recent years have seen substantial progress in parametric and nonlinear model reduction for large-scale systems. Methods for linear time-invariant systems are by now well established and include the proper orthogonal decomposition (POD) (Berkooz et al., 1993; Holmes et al., 1996; Sirovich, 1987), Krylov-based methods (Feldmann and Freund, 1995; Gallivan et al., 1994), balanced truncation (Moore, 1981), and reduced-basis methods (Noor and Peters, 1980; Ghanem and Sarkar, 2003). Extensions of these methods to handle nonlinear and parametrically varying problems have played a major role in moving model reduction from forward simulation and control to applications in optimization and UQ.

Several methods have been developed for nonlinear model reduction. One approach is to use the trajectory piecewise-linear scheme, which employs a weighted combination of linear models, obtained by linearizing the nonlinear system at selected points along a state or parameter trajectory (Rewienski and White, 2003). Other approaches propose using a reduced basis or POD model-reduction approach and approximating the nonlinear term through the selective sampling of a subset of the original equations (Bos et al., 2004; Astrid et al., 2008; Barrault et al., 2004; Grepl et al., 2007). For example, in Astrid et al. (2008), the missing-point-estimation approach, based on the theory of "gappy POD" (Everson and Sirovich, 1995), is used to approximate nonlinear terms in the reduced model with selective spatial sampling. The empirical interpolation method (EIM) is used to approximate the nonlinear terms by a linear combination of empirical basis functions for which the coefficients are determined using interpolation (Barrault et al., 2004; Grepl et al., 2007). Recent work has established the discrete empirical interpolation method (DEIM) (Chaturantabut and Sorensen, 2010), which extends the EIM to a more general class of problems. Although these methods have been successful in a range of applications, several challenges remain for nonlinear model reduction. For example, current methods pose limitations on the form of the nonlinear system that can be considered, and problems with nonlocal nonlinearities can be challenging.

For parametric model reduction, several classes of methods have emerged. Each approach handles parametric variation in a different way, although there is a common theme of interpolation among information collected at different parameter values. The EIM and DEIM methods described above can be used to handle some classes of parametrically varying systems. In the circuit community, Krylov-based methods have been extended to include parametric variations, again with a restriction on the form of systems that can be considered (Daniel et al., 2004). Another class of approaches approximates the variation of the projection basis as a function of the parameters (Allen et al., 2004; Weickum et al., 2006). An alternative for expansion of the basis is interpolation among the reduced subspaces (Amsallem et al., 2007)—for example, using interpolation in the space tangent to the Grassmannian manifold of a POD basis constructed at different parameter points—or among the reduced models (Degroote et al., 2010).

Historically, the use of reduced-order models in UQ is not as common as the surrogate modeling methods (e.g., see Section 4.2) that approximate the full inputs-to-observable map. Some recent examples of model reduction for UQ include statistical inverse problems (Lieberman et al., 2010; Galbally et al., 2010), forward propagation of uncertainty for heat conduction (Boyaval et al., 2009), computational fluid dynamics (Bui-Thanh and Wilcox, 2008), and materials (Kouchmeshky and Zabaras, 2010).

Finding: An important area of future work is the use of model reduction for optimization under uncertainty.

4.2 FORWARD PROPAGATION OF INPUT UNCERTAINTY

In many settings, inputs to a computer model are uncertain and therefore can be treated as random variables. Interest then lies in propagating the uncertainty in the input distribution to the output of the deterministic computer model. It is the distribution of the model outputs, or some feature thereof, that is the primary interest of a typical investigation. For example, one may be interested in the 95th percentile of the output distribution or, perhaps, the mean of the output, along with associated uncertainties.

The set of approaches that either treat the computational model as a black box (nonintrusive techniques) or require modifications to the underlying mathematical model (intrusive techniques) is considered here. In addition, the activities are separated into settings in which (1) the number of simulator evaluations is essentially unlimited (e.g., Monte Carlo and polynomial chaos) and (2) only a relatively small number of simulator runs is available (e.g., Gaussian process, polynomial chaos, and quasi-Monte Carlo).

In a most straightforward manner Monte Carlo sampling can be used—where one can sample directly from the input distribution and evaluate the output of the computer model at each input setting. The estimates for the QOI (mean response, confidence intervals, percentiles, and so on) are obtained from the induced empirical distribution function of the model outputs. Generally, Monte Carlo sampling does not depend on the dimension of the input space or on the complexity of the model. This makes Monte Carlo sampling an attractive approach when the forward model is complicated, has many inputs, and is sufficiently fast. However, in many real-world problems, Monte Carlo methods can require thousands of code executions to produce sufficient accuracy. The number of required function evaluations can be significantly reduced by using quasi-Monte Carlo methods (e.g., see Lemieux, 2009). These rely on relatively few input configurations and attempt to mimic the properties of a random sample to estimate features of the output distribution (e.g., Latin hypercube sampling; McKay et al., 1979; Owen, 1997).

Monte Carlo-based approaches are ill-equipped to take advantage of physical or mathematical structure that could otherwise expedite the calculations. In its most fundamental form, sampling does not retain the functional dependence of output on input but rather produces quantities that have been averaged simultaneously over all input parameters. In contradistinction, the polynomial chaos (PC) methodology (Ghanem and Spanos, 1991; Soize and Ghanem, 2004; Najm, 2009; Xiu and Karniadakis, 2002; Xiu, 2010), capitalizes on the mathematical structure provided by the probability measures on input parameters to develop approximation schemes with a priori convergence results inherited from Galerkin projections and spectral approximations common in numerical analysis. The PC methodology has two essential components. A first step involves a description of stochastic functions, variables, or processes with respect to a basis in a suitable vector space. The choice of basis can be adapted to the distribution of the input parameters (e.g., Xiu and Karniadakis, 2002; Soize and Ghanem, 2004). The second step involves computing the coordinates in this representation using functional analysis machinery, such as orthogonal projections and error minimization. These coordinates permit the rapid evaluation of output quantities of interest as well as of the sensitivities (both local and global) of output uncertainty with respect to input uncertainty.

Two procedures have generally pursued the connection to PC approximations: intrusive and nonintrusive. The so-called nonintrusive approach computes the coefficients in the PC decomposition as multidimensional integrals that are approximated by means of sparse quadrature and other numerical integration rules. The intrusive approach, however, synthesizes new equations that govern the behavior of the coefficients in the PC decomposition from the initial governing equations. Intrusive uncertainty propagation approaches involve a reformulation of the forward model, or its adjoint, to directly produce a probabilistic representation of uncertain model outputs induced by input uncertainty. Such approaches include PC methods, as well as extensions of local sensitivity analysis and adjoint methods. In either case, unlike the first type of emulators described in Section 4.1.1, this class of emulator aims to directly map uncertainty of model input to uncertainty of model output.

Another strategy is to first construct an emulator of the computer model (see Section 4.1.1) and then propagate the distribution of inputs through the emulator (Oakley and O'Hagan, 2002; Oakley, 2004; Cannamela et al., 2008). Essentially, one treats the emulator as if it were a fast computer model and uses the methods already discussed. In these cases, one must account for the variability in the output distribution induced by the random variable as well as the uncertainty in emulating the computer model (Oakley, 2004). Of course, almost all approaches, outside of straight Monte Carlo, will be faced with the curse of dimensionality. Accounting for this induced variability is an important direction for future research. Interestingly, if the computer model can be assumed to be a GP with known correlation/variance parameters, the propagation of the uncertainty distribution of the model inputs across the GP (i.e., the response surface approach) and polynomial chaos can be viewed as alternative approaches to the same problem, both giving effective approaches for exploring the uncertainty in QOIs.

Consider the particularly challenging problem described in Section 4.5 in which interest lies in estimating statistics (e.g., mean and standard deviation) for observables in an electromagnetic interference (EMI) application. The solution combines features of polynomial chaos expansions and also of emulation followed by uncertainty propagation (Oakley, 2004). For this application, the computer model outputs behave quasi-chaotically as a function of the input configurations. In such cases, emulators that leverage local information instead of a single global model are often more effective. In this specific case study, the computer model is emulated by a nonintrusively computed polynomial chaos decomposition and uses a robust emulator that attempts to model local features. The approach taken is similar to that of Oakley (2004) insofar as the emulator is first constructed and then the distribution of QOIs is approximated by propagating the distribution of inputs across the emulator.

These methods for propagating the variation of the input distribution to explore the uncertainty in the model output are particularly effective when the QOIs are estimates that are centrally located in the output distribution. However, decision makers are often most interested in rare events (e.g., extreme weather scenarios, or a combination of conditions causing system failure). These cases are located in the tails of the output distribution where exploration is often impractical with the aforementioned methods. This problem is only exacerbated for high-dimensional inputs. An important research direction is the estimating of the probability of rare events in light of complicated models and input distributions. One approach to this problem is to bias the model output toward these rare events and properly account for this biasing. Importance sampling (Shahabuddin, 1994) is another approach.

Finding: Further research is needed to develop methods to identify the input configurations for which a model predicts significant rare events and for assessing their probabilities.

4.3 SENSITIVITY ANALYSIS

Sensitivity analysis (SA) is concerned with understanding how changes in the computational-model inputs influence the outputs or functions of the outputs. There are several motivations for SA, including the following: enhancing the understanding of a complex model, finding aberrant model behavior, seeking out which inputs have a substantial effect on a particular output, exploring how combinations of inputs interact to affect outputs, seeking out regions of the input space that lead to rapid changes or extreme values in output, and gaining insight as to what additional information will improve the model's ability to predict. Even when a computational model is not adequate for reproducing physical system behavior, its sensitivities may still be useful for developing inferences about key features of the physical system.

In many cases, the most general and physically accurate computational model takes too long to run, making impractical its use as the main tool for a particular investigation and forcing one to seek a simpler, less computationally demanding model. SA can serve as a first step in constructing an emulator and/or a reduced model of a physical system, capturing the important features in a large-scale computational model while sacrificing complexity to speed up the run time. A perfunctory SA also serves as a simplest first step to characterizing output uncertainty induced by uncertainty in inputs (uncertainty analysis is discussed below in this chapter).

Implicit in SA is an underlying surrogate that permits the efficient mapping of changes in input to changes in output. This surrogate highlights the value of emulators and reduced-order models to SA. A few cases in which the peculiar structure of the surrogates allows the analytical evaluation of key sensitivity quantities have attracted particular attention and served to shape the current practice of SA. Local SA is associated with linearization of the input-output map at some judiciously chosen points. Higher-order Taylor expansions have also been used to enhance the accuracy of these surrogates. Global SA, however, relies on global surrogates that better capture the effects of interactions between random variables. Two particular forms of global surrogates have been pursued in recent years that rely, respectively, on polynomial chaos decompositions and Sobol's decomposition. These decompositions permit the computation of the sensitivity of output variance with respect to the variances of the individual

EMULATION, REDUCED-ORDER MODELING, AND FORWARD PROPAGATION

input parameters. Given the dominance in current research and practice of these particular interpretations of global and local sensitivity, these global SA methods and local SA methods are detailed in the remainder of this section.

4.3.1 Global Sensitivity Analysis

Global SA seeks to understand a complex function over a broad space of input settings, decomposing this function into a sum of increasingly complex components (see Box 4.1). These components might be estimated directly (Oakley and O'Hagan, 2004), or each of these components can be summarized by a variance measure

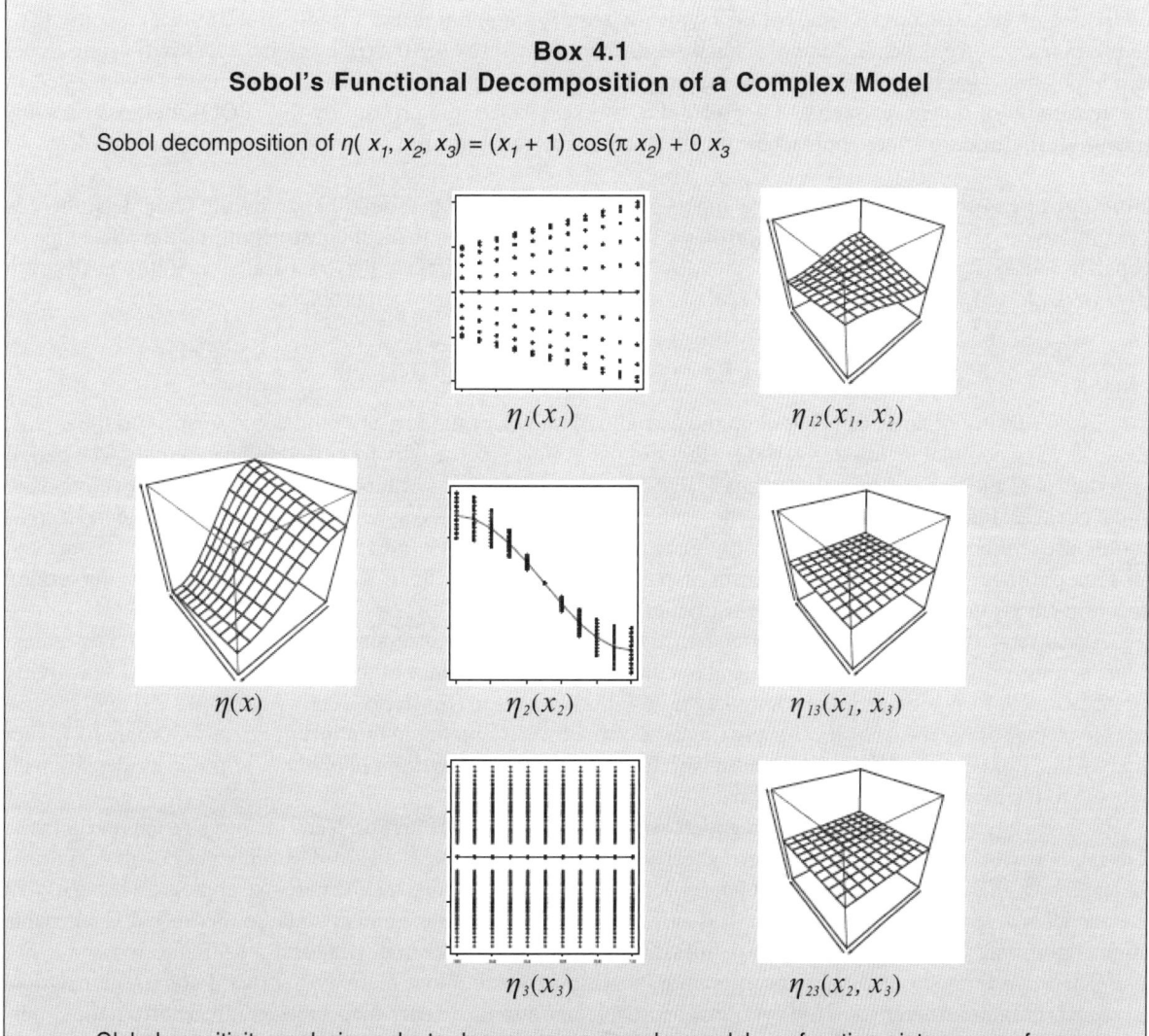

**Box 4.1
Sobol's Functional Decomposition of a Complex Model**

Sobol decomposition of $\eta(x_1, x_2, x_3) = (x_1 + 1)\cos(\pi x_2) + 0\, x_3$

$\eta_1(x_1)$ $\eta_{12}(x_1, x_2)$

$\eta(x)$ $\eta_2(x_2)$ $\eta_{13}(x_1, x_3)$

$\eta_3(x_3)$ $\eta_{23}(x_2, x_3)$

Global sensitivity analysis seeks to decompose a complex model—or function—into a sum of a constant plus main effects plus interactions (Sobol, 1993). Here a three-dimensional function is decomposed into three main effect functions (red lines) η_1, η_2, and η_3 plus three two-way interaction functions η_{12}, η_{13}, and η_{23} plus a single three-way interaction (not shown). Global sensitivity analysis seeks to estimate these component functions or variance measures of these component functions. See Saltelli et al. (2000) or Oakley and O'Hagan (2004) for examples.

estimated using one of a variety of techniques: Monte Carlo (MC) (Kleijnen and Helton, 1999); regression (Helton and Davis, 2000); analysis of variance (Moore and McKay, 2002); Fourier methods (Sobol, 1993; Saltelli et al., 2000); or GP or other response-surface models (Oakley and O'Hagan, 2004; Marzouk and Najm, 2009). Each of these approaches requires an ensemble of forward model runs to be carried out over a specified set of input settings.

The number of computational-model runs required will depend on the complexity of the forward model over the input space and the dimension of the input space, as well as the estimation method. A key challenge for global SA—indeed, for much of VVUQ—is to carry out such analyses with limited computational resources. The various estimation methods deal with this issue in one way or another. For example, MC methods, while generally requiring many model runs to gain reasonable accuracy, can handle high-dimensional input spaces and arbitrary complexity in the computational model. In contrast, response-surface-based approaches, such as GP, may use an ensemble consisting of very few model runs, but they typically require smoothness and sparsity in the model response (i.e., the response surface depends on only a small set of the inputs). Box 4.1 describes global sensitivity using Sobol decompositions—although these are not the only sensitivity measures. For example, one may be interested in the sensitivity of quantities such as the probability of exceedance (the probability that a QOI will exceed some quantity). In this case, other approaches to global sensitivity must be used.

Finding: In cases where a large-scale computational model is built up from a hierarchy of submodels, there is opportunity to develop efficient SA approaches that leverage this hierarchical construction, taking advantage of separate sensitivity analyses on these submodels. Exactly how to aggregate these separate sensitivity anslyses to give accurate sensitivities on the larger-scale model outputs remains an open problem.

4.3.2 Local Sensitivity Analysis

Local sensitivity analysis is based on the partial derivatives of the forward model with respect to the inputs, evaluated at a nominal input setting. Hence, this sort of SA makes sense only for differentiable outputs. The partial derivatives of the forward model, perhaps scaled, can serve as some indication of how model output responds to input changes. First-order sensitivities—the gradient of an output of interest with respect to inputs—are most commonly used, although second-order sensitivities incorporating partial or full Hessian information can be obtained in some settings. Even higher-order sensitivities have been pursued, albeit with increasing difficulty (emerging tensor methods may make third-order sensitivities more feasible).

Local sensitivities give limited information about forward-model behavior in the neighborhood of the nominal input setting, providing some information about the input-output response of the forward model. Local sensitivity, or derivative information, is more commonly used for optimization (Finsterle, 2006; Flath et al., 2011) in inverse problems (see Section 4.4) or local approximation of the forward model in nonlinear regression problems (Seber and Wild, 2003). In these cases, it is the sensitivity of an objective function, likelihood, or posterior density with respect to the model inputs that is required.

There are two approaches to obtaining local sensitivities: a black box approach and an intrusive approach. In the black box approach, the underlying mathematical or computational model is regarded as being inaccessible, as might be the case with an older, established, or poorly documented code. By contrast, the intrusive approach presumes that the model is accessible, whether because it is sufficiently well documented and modular, or because it is amenable to the retrofitting of local sensitivity capabilities, or because it was developed with local sensitivities in mind.

There are limited options for incorporating local sensitivities into a black box forward code. The classical approach is finite differences. However, this approach can provide highly inaccurate gradient information, particularly when the underlying forward model is highly nonlinear, and "solving" the forward problem amounts to being content with reducing the residual by several orders of magnitude. Moreover, the cost of finite differencing grows linearly with the number of inputs.

An alternative approach is provided by automatic differentiation (AD), sometimes known as algorithmic differentiation. Assuming that one has access to source code, AD is able to produce sensitivity information directly from the source code by exploiting the fact that a code is written from a series of elementary operations, whose known derivatives can be chained together to provide exact sensitivity information. This approach avoids the numerical difficulties of finite differencing. Furthermore, the so-called reverse mode of AD can be employed to generate gradient information at a cost that is independent (to leading order) of the number of inputs. However, the basic difficulty with AD methods is that they differentiate the code rather than the underlying mathematical model. For example, while the sensitivity equations are linear, AD differentiates through the nonlinear solver; and while sensitivity equations share the same operator, AD repeatedly differentiates through the preconditioner. This also means that artifacts of the discretization (e.g., adaptive mesh refinement) are differentiated. Additionally, when the code is large and complex, current AD tools will often break down. Still, when they do work, and when the intrusive methods described below are not feasible because of time constraints or lack of modularity of the forward code, AD can be a viable approach.

If one has access to the underlying forward model, or if one is developing a local sensitivity capability from scratch, one can overcome many of the difficulties outlined above by using an intrusive method. These methods differentiate the mathematical model underlying the code. This can be done at the continuum level, yielding mathematical derivatives that can then be discretized to produce numerical derivatives. Alternatively, the discretized model may be directly differentiated. These two approaches do not always result in the same derivatives, though often (e.g., with Galerkin discretization) they do. The advantage of differentiating the underlying mathematical model is that one can exploit the model structure. The equations that govern the derivatives of the state variables with respect to each parameter—the so-called sensitivity equations—are linear, even when the forward problem is nonlinear, and they are characterized by the same coefficient matrix (or operator) for each input. This matrix is the Jacobian of the forward model, and thus Newton-based forward solvers can be repurposed to solve the sensitivity equations. Because each right-hand side corresponds to the derivative of the model residual with respect to each parameter, the construction of the preconditioner can be amortized over all of the inputs.

Still, solving the sensitivity equations may prove too costly when there are large numbers of inputs. An alternative to the sensitivity equation approach is the so-called adjoint approach, in which an adjoint equation[1] is solved for each output of interest, and the resulting adjoint solution is used to construct the gradient (analogs exist for higher derivatives). This adjoint equation is, like the sensitivity equation, always linear in the adjoint variable, and its right-hand side corresponds to the derivative of the output with respect to the state. Its operator is the adjoint (or transpose) of the linearized forward model, and so here again preconditioner construction can be amortized, in this case over the outputs of interest. When the number of outputs is substantially less than the number of input parameters, the adjoint approach can result in substantial savings. By postponing discretization until after derivatives have been derived (through variational means), one can avoid differentiating artifacts of the discretization (e.g., subgrid-scale models, flux limiters).

Even first derivatives can greatly extend one's ability to explore the forward model's behavior in the service of UQ, especially when the input dimension is large. If derivative information can be calculated for the cost of only an additional model run, as is often the case with adjoint models, then tasks such as global SA, solving inverse problems, and sampling from a high-dimensional posterior distribution can be carried out with far less computational effort, making currently intractable problems tractable.

Generalizing adjoint methods to better tackle difficult computational problems such as multiphysics applications, operator splitting, and nondifferentiable solution features (such as shocks) would extend the universe of problems for which derivative information can be efficiently computed. At the same time, developing and extending UQ methods to take better advantage of derivative information will broaden the universe of problems for which computationally intensive UQ can be carried out. Current examples of UQ methods applicable to large-scale computational models that take advantage of derivative information include normal linear approximations for inverse problems (Cacuci et al., 2005), response surface methods (Mitchell et al., 1994), and Markov chain MC sampling techniques for Bayesian inverse problems (Neal, 1993; Girolami and Calderhead, 2011).

Finding: There is potential for significant benefit from research and development in the compound area of (1) extracting derivatives and other features from large-scale computational models and (2) developing UQ methods that efficiently use this information.

---

[1] A discussion is given in Marchuk (1995).

4.4 CHOOSING INPUT SETTINGS FOR ENSEMBLES OF COMPUTER RUNS

An important decision in an exploration of the simulation model is the choice of simulation runs (i.e., experimental design) to perform. Ultimately, the task at hand is to provide an estimate of some feature of the computer model response as efficiently as possible. As one would expect, the optimal set of model evaluations is related to the specific aims of the experimenter.

For physical experiments, there are three broad principles for experimental design: randomization, replication, and blocking.[2] For deterministic computer experiments, these issues do not apply—replication, for example, is just wasted effort. In the absence of prior knowledge of the shape of the response surface, however, a simple rule of thumb worth following is that the design points should be spread out to explore as much of the input region as possible.

Current practice for the design of computer experiments identifies strategies for a variety of objectives. If the goal is to identify the active factors governing the system response, one-at-a-time designs[3] are commonly used (Morris, 1991). For computer-model emulation, space-filling designs (Johnson and Schneiderman, 1991), and Latin hypercube designs (McKay et al., 1979) and variants thereof (Tang, 1993; Lin et al., 2010) are good choices. The designs used for building emulators are often motivated by space-filling and projection properties that are important for quasi-MC studies (Lemieux, 2009). In studies in which the goal is to estimate a feature of the computer-model response surface, such as a global maximum or level sets, sequential designs have proven effective (Ranjan et al., 2008). In these cases, one is usually attempting to select new simulator trials that aim to optimally improve the estimate of the feature of interest rather than the estimate of the entire response surface. For SA, common design strategies include fractional factorial and response-surface designs (Box and Draper, 2007), as well as Morris's one-at-a-time designs and other screening designs (e.g., Saltelli and Sobol, 1995). The use of adaptive strategies for the larger V&V problem is discussed in Chapter 5.

4.5 ELECTROMAGNETIC INTERFERENCE IN A TIRE PRESSURE SENSOR: CASE STUDY

4.5.1 Background

The proper functioning of electronic communication, navigation, and sensing systems often is disturbed, upset, or blocked altogether by electromagnetic interference (EMI)—viz, natural or man-made signals that are foreign to the systems' normal mode of operation. Natural EMI sources include atmospheric charge/discharge phenomena such as lightning and precipitation static. Man-made EMI can be intentional—arising from jamming or electronic warfare—or unintentional, resulting from spurious electromagnetic emissions originating from other electronic systems.

To guard mission-critical electronic systems against interference and ensure system interoperability and compatibility, engineers employ a variety of electromagnetic shielding and layout strategies to prevent spurious radiation from penetrating into, or escaping from, the system. This practice is especially relevant for consumer electronics subject to regulation by the Federal Communications Commission. In developing EMI mitigation strategies, it is important to recognize that many EMI phenomena are stochastic in nature. The degree to which EMI affects a system's performance is influenced by its electromagnetic environment, e.g., its mounting platform and proximity to natural or man-made sources of radiation. Unfortunately, a system's electromagnetic environment oftentimes is ill-characterized at the time of design. The uncertainty in the effect of EMI on a system's performance is further exacerbated by variability in its electrical and material component values and geometric dimensions.

4.5.2 The Computer Model

Although the EMI compliance of a system prior to deployment or mass production always is verified experimentally, engineers increasingly rely on modeling and simulation to reduce costs associated with the building and

---
[2] Blocking is the arrangement of experimental units into groups.
[3] One-at-a-time designs are designs that vary one variable at a time.

testing of prototypes early in the design process. EMI phenomena are governed by Maxwell's equations. These equations have astonishing predictive power, and their reach extends far beyond EMI analysis. Indeed, Maxwell's equations form the foundation of electrodynamics and classical optics and the underpinnings of many electrical, computer, and communication technologies. Fueled by advances in both algorithms and computer hardware, Maxwell equation solvers have become indispensable in scientific and engineering disciplines ranging from remote sensing and biomedical imaging to antenna and circuit design, to name but a few.

The application of VVUQ concepts to the statistical characterization of EMI phenomena described below leverages an integral equation-based Maxwell equation solver. The solver accepts as input a computer-aided design (CAD) description of a system's geometry along with its external excitation, and returns a finite-element approximation of the electrical currents on the system's conducting surfaces, shielding enclosures, printed circuit boards, wires/cables, and dielectric (plastic) volumes (Bagci et al., 2007). To enable the simulation of EMI phenomena on large- and multiscale computing platforms, the solver executes in parallel and leverages fast and highly accurate O[N log(N)] convolution methods, causing its computational cost to scale roughly linearly with the number of unknowns in the finite-element expansion. To facilitate the characterization of real-world EMI phenomena, the solver interfaces with a Simulation Program with Integrated Circuit Emphasis (SPICE)[4]-based circuit solver that computes node voltages on lumped element circuits that model electrically small components. Finally, to allow for the characterization of a "system of systems," the Maxwell equation solver interfaces with a cable solver that computes transmission-line voltages and currents on transmission lines interconnecting electronic (sub)systems (Figure 4.2).

The application of this hybrid analysis framework to the statistical characterization of EMI phenomena in real-world electronic systems is illustrated by means of a tire pressure monitoring (TPM) system (Figure 4.3). TPM systems monitor the air pressure of vehicle tires and warn drivers when a tire is under-inflated. The most widely used TPM system uses a small battery-operated sensor-transponder mounted on a car's tire rim just behind the valve stem. The sensor-transponder transmits information on the tire pressure and temperature to a central TPM

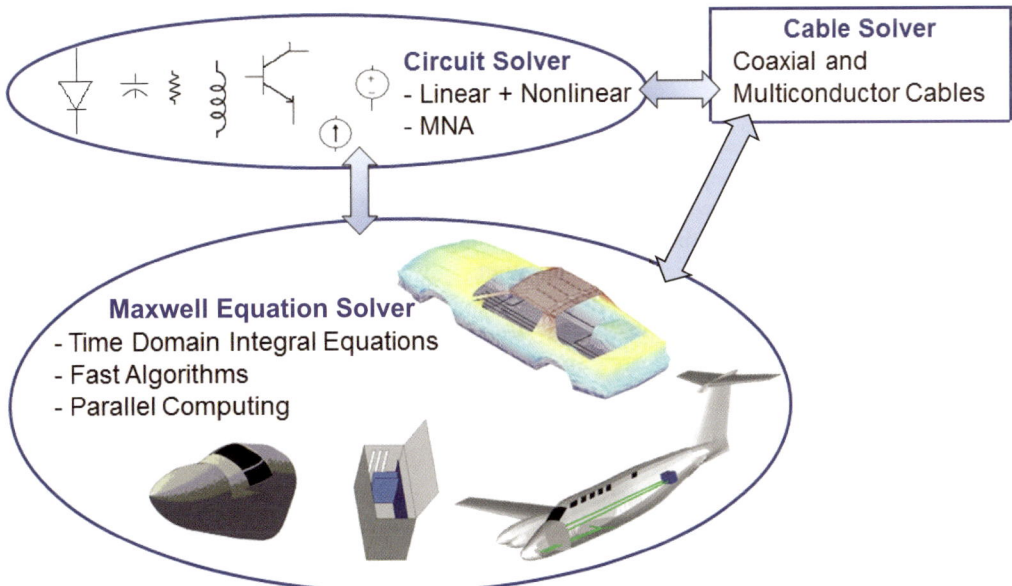

FIGURE 4.2. Hybrid electromagnetic interference analysis framework composed of Maxwell equation, circuit, and cable solvers.

---

[4] SPICE is a general-purpose open-source analog electronic circuit simulator.

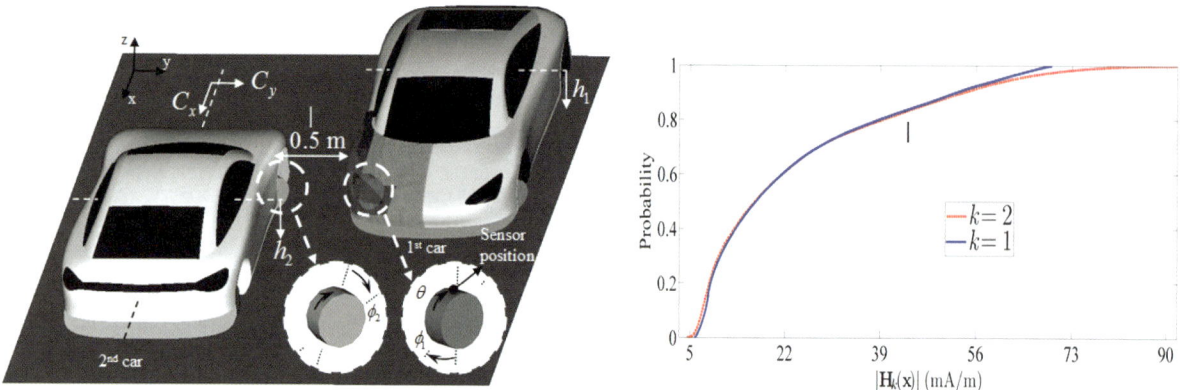

FIGURE 4.3 (*Left*) Two cars with tire pressure monitoring (TPM) systems mounted on their front-passenger-side wheel rims. (*Right*) Comparison of cumulative distribution functions of the TPM received signal on one car, without (k = 1) and with (k = 2) the second car present.

receiver mounted on the body of the car. In this case study, the strength of the received signal when the system is subject to EMI originating from another nearby car using the same system is characterized. The received signal depends on the relative position of the two cars, described by seven parameters: the rotation and steering angles of the wheels of both cars carrying the TPM transponders, the height of the car bodies with respect to their wheel base, and the relative position of the cars with respect to each other.

In principle, statistics pertaining to the strength of the received signal can be deduced by an MC method, namely by repeatedly executing the Maxwell equation solver for many realizations of the random parameters sampled with respect to their probability distribution functions, which here are assumed to be uniform. Unfortunately, while such an MC method is straightforward to implement, for the problem at hand it requires hundreds of thousands of deterministic code executions to converge. The slow convergence of the MC method combined with the fact that execution of the deterministic Maxwell equation solver for the TPM problem requires roughly 1 hour of central processing unit (CPU) time all but rules out its direct application.

4.5.3 Robust Emulators

To avoid the pitfalls associated with the direct application of MC, an emulator (or surrogate model) for the strength of the received TPM signal (the QOI) as a function of the seven input parameters was constructed for this study. The emulator provides an accurate approximation of the received signal for all combinations of the parameters, yet can be evaluated in a fraction of the time required for executing the Maxwell equation solver. The emulator thus enables the cost-effective, albeit indirect, application of MC to the statistical characterization of the received TPM signal. In this case study, the emulator was constructed by means of a multielement stochastic collocation (ME-SC) technique that leverages generalized polynomial chaos (gPC) expansions to represent the received signal (Xiu, 2007; Agarwal and Aluru, 2009).

The ME-SC method is an extension of the basic SC method, which approximates a selected QOI (in this case the received TPM signal) by polynomials spanning the entire input parameter space. The SC method constructs these polynomials by calling a deterministic simulator—here a Maxwell solver—to evaluate the QOI for combinations of random inputs specified by collocation points in the seven-dimensional input space. Unfortunately, basic SC methods oftentimes become impractical and inaccurate for outputs that vary rapidly or nonsmoothly with changes in the input parameters, because their representation calls for high-order polynomials. Such is the case in EMI analysis, in which voltages across system pins and currents on circuit traces and the received TPM signal strength, behave rapidly and sometimes quasi-chaotically in the input space. Fortunately, extensions to SC methods have been developed that remain efficient and accurate for modeled outputs with nonsmooth and/or discontinuous dependencies on the inputs. The ME-SC method is one of them. It achieves its efficiency and accuracy by adaptively dividing the input space into subdomains based on the decay rates of the outputs' local variances and constructing separate polynomial approximations for each subdomain (Agarwal and Aluru, 2009).

The use of emulators adds uncertainty to the process of statistically characterizing EMI phenomena for which it is often difficult to account. Indeed, the construction of the emulator by means of ME-PC methods involves a greedy search for a sparse basis for the random inputs. When applied to complex, real-world problems, this search is not guaranteed to converge or to yield an accurate representation of the QOI. In the context of EMI analysis, emulator techniques often are applied to simple toy problems that qualitatively relate to the real-world problem at hand, yet allow for an exhaustive canvasing of input space. If and when the method performs well on the toy problem, it is then applied to more complex, real-world scenarios, often without looking back.

4.5.4 Representative Result

The ME-SC emulator models the signal strength in the TPM receiver in one car radiated by simultaneously active sensor-transponders in both cars. Construction of the ME-SC emulator required 545 calls of the Maxwell equation solver, a very small fraction of the number of calls required in the direct application of MC. The relative accuracy of the ME-SC emulator with respect to the signal strength predicted by the Maxwell equation solver was below 0.1 percent for each of the 545 system configurations in the seven-dimensional input space. The cost of applying MC to the emulator is negligible compared to that of a single call to the Maxwell equation solver. Figure 4.3(right) shows the cumulative distribution function (cdf) of the received signal (QOI) and compares it to the cdf produced with only one car present. The presence of the second car does not substantially alter the cdf for small values of the received signal. It does substantially increase the maximum possible received signal, albeit not enough to cause system malfunction.

4.6 REFERENCES

Agarwal, N., and N. Aluru. 2009. A Domain Adaptive Stochastic Collocation Approach for Analysis of MEMS Under Uncertainties. *Journal of Computational Physics* 228:7662-7688.

Allen, M., G. Weickum, and K. Maute. 2004. Application of Reduced-Order Models for the Stochastic Design Optimization of Dynamic Systems. Pp. 1-20 in *Proceedings of the 10th AIAA/ISSMO Multidisciplinary Optimization Conference*. August 30-September 1, 2004. Albany, N.Y.

Amsallem, D., C. Farhat, and T. Lieu. 2007. High-Order Interpolation of Reduced-Order Models for Near Real-Time Aeroelastic Prediction. Paper IF-081:18-20. International Forum on Aeroelasticity and Structural Dynamics. Stockholm, Sweden.

Astrid, P., S. Weiland, K. Willcox, and T. Backx. 2008. Missing Point Estimation in Models Described by Proper Orthogonal Decomposition. *IEEE Transactions on Automatic Control* 53(10):2237-2251.

Bagci, H., A.E. Yilmaz, J.M. Jin, and E. Michielssen. 2007. Fast and Rigorous Analysis of EMC/EMI Phenomena on Electrically Large and Complex Cable-Loaded Structures. *IEEE Transactions on Electromagnetic Compatibility* 49:361-381.

Barrault, M., Y. Maday, N.C. Nguyen, and A.T. Patera. 2004. An "Empirical Interpolation" Method: Application to Efficient Reduced-Basis Discretization of Partial Differential Equations. *Comptes Rendus Mathematique* 339(9):667-672.

Ben-Ari, E.N., and D.M. Steinberg. 2007. Modeling Data from Computer Experiments: An Empirical Comparison of Kriging with MARS and Projection Pursuit Regression. *Quality Engineering* 19:327-338.

Berkooz, G., P. Holmes, and J.L. Lumley. 1993. The Proper Orthogonal Decomposition in the Analysis of Turbulent Flows. *Annual Review of Fluid Dynamics* 25:539-575.

Bos, R., X. Bombois, and P. Van den Hof. 2004. Accelerating Large-Scale Non-Linear Models for Monitoring and Control Using Spatial and Temporal Correlations. *Proceedings of the American Control Conference* 4:3705-3710.

Box, G.E.P., and N.R. Draper. 2007. *Response Surfaces, Mixtures, and Ridge Analysis*. Wiley Series in Probability and Statistics, Vol. 527. Hoboken, N.J.: Wiley.

Boyaval, S., C. LeBris, Y. Maday, N.C. Nguyen, and A.T. Patera. 2009. A Reduced Basis Approach for Variational Problems with Stochastic Parameters: Application to Heat Conduction with Variable Robin Coefficient. *Computer Methods in Applied Mechanics and Engineering* 198:3187-3206.

Bui-Thanh, T., and K. Willcox. 2008. Parametric Reduced-Order Models for Probabilistic Analysis of Unsteady Aerodynamic Applications. *AIAA Journal* 46 (10):2520-2529.

Cacuci, D.C., M. Ionescu-Bujor, and I.M. Navon. 2005. *Sensitivity and Uncertainty Analysis: Applications to Large-Scale Systems*, Vol. 2. Boca Raton, Fla.: CRC Press.

Cannamela, C., J. Garnier, and B. Iooss. 2008. Controlled Stratification for Quantile Estimation. *Annals of Applied Statistics* 2(4):1554-1580.

Chaturantabut, S., and D.C. Sorensen. 2010. Nonlinear Model Reduction via Discrete Empirical Interpolation. *SIAM Journal on Scientific Computing* 32(5):2737-2764.

Clarke, S.M., M.D. Zaeh, and J.H. Griebsch. 2003. Predicting Haptic Data with Support Vector Regression for Telepresence Applications. Pp. 572-578 in *Design and Application of Hybrid Intelligent Systems*. Amsterdam, Netherlands: IOS Press.

Daniel, L., O.C. Siong, L.S. Chay, K.H. Lee, and J. White. 2004. A Multiparameter Moment-Matching Model-Reduction Approach for Generating Geometrically Parameterized Interconnect Performance Models. *IEEE Transactions on Computer-Aided Design of Integrated Circuits and Systems* 23(5):678-693.

Degroote, J., J. Vierendeels, and K. Wilcox. 2010. Interpolation Among Reduced-Order Matrices to Obtain Parametrized Models for Design, Optimization and Probabilistic Analysis. *International Journal for Numerical Methods in Fluids* 63(2):207-230.

Everson, R., and L. Sirovich. 1995. Karhunen-Loève Procedure for Gappy Data. *Journal of the Optical Society of America* 12(8):1657-1664.

Feldman, P., and R.W. Freund. 1995. Efficient Linear Circuit Analysis by Pade Appoximation via the Lanczos Process. *IEEE Transactions on Computer-Aided Design of Integrated Circuits* 14(5):639-649.

Finsterle, S. 2006. Demonstration of Optimization Techniques for Groundwater Plume Remediation Using Itough2. *Environmental Modeling and Software* 21(5):665-680.

Flath, H.P., L.C. Filcox, V. Akcelik, J. Hill, B. Van Bloeman Waanders, and O. Glattas. 2011. Fast Algorithms for Bayesian Uncertainty Quantification in Large-Scale Linear Inverse Problems Based on Low-Rank Partial Hessian Approximations. *SIAM Journal on Scientific Computing* 33(1):407-432.

Floater, M.S., and A. Iske. 1996. Multistep Scattered Data Interpolation Using Compactly Supported Radial Basis Functions. *Journal of Computational and Applied Mathematics* 73(1-2):65-78.

Galbally, D., K. Fidkowski, K. Willcox, and O. Ghattas. 2010. Non-Linear Model Reduction for Uncertainty Quantification in Large-Scale Inverse Problems. *International Journal for Numerical Methods in Engineering* 81:1581-1608.

Gallivan, K.E., E. Grimme, and P. Van Dooren. 1994. Pade Approximations of Large-Scale Dynamic Systems with Lanczos Methods. *Decision and Control: Proceedings of the 33rd IEEE Conference* 1:443-448.

Ghanem, R., and A. Sarkar. 2003. Reduced Models for the Medium-Frequency Dynamics of Stochastic Systems. *Journal of the Acoustical Society of America* 113(2):834-846.

Ghanem, R., and P. Spanos. 1991. A Spectral Stochastic Finite Element Formulation for Reliability Analysis. *Journal of Engineering Mechanics, ASCE* 117(10):2351-2372.

Girolami, M., and B. Calderhead. 2011. Reimann Manifold Langevin and Hamiltonian Monte Carlo Methods. *Journal of the Royal Society: Series B (Statistical Methodology)* 73(2):123-214.

Gramacy, R.B., and H.K.H. Lee. 2008. Bayesian Treed Gaussian Process Models with an Application to Computer Modeling. *Journal of the American Statistical Association* 103(483):1119-1130.

Grepl, M.A., Y. Maday, N.C. Nguyen, and A.T. Patera. 2007. Efficient Reduced-Basis Treatment of Nonaffine and Nonlinear Partial Differential Equations. *ESAIM-Mathematical Modeling and Numerical Analysis (M2AN)* 41:575-605.

Hayken, S. 1998. *Neural Networks: A Comprehensive Foundation*. Upper Saddle River, N.J.: Prentice Hall.

Helton, J.C., and F.J. Davis. 2000. *Sampling-Based Methods for Uncertainty and Sensitivity Analysis*. Albuquerque, N.Mex.: Sandia National Laboratories.

Holmes, P., J.L. Lumley, and G. Berkooz. 1996. *Turbulence, Coherent Structures, Dynamical Systems, and Symmetry*. Cambridge, U.K.: Cambridge University Press.

Iooss, B., and M. Ribatet. 2007. Global Sensitivity Analysis of Stochastic Computer Models with Generalized Additive Models. *Technometrics*.

Jin, R., W. Chen, and T.W. Simpson. 2000. Comparative Studies of Metamodeling Techniques. *European Journal of Operations Research* 138:132-154.

Johnson, B., and B. Schneiderman. 1991. Tree-Maps: A Space-Filling Approach to the Visualization of Hierarchical Information Structures. Pp. 284-291 in *Proceedings of IEEE Conference on Visualization*.

Kleijnen, J.P.C., and J.C. Helton. 1999. Statistical Analyses of Scatterplots to Identify Important Factors in Large-Scale Simulations. 1: Review and Comparison of Techniques. *Reliability Engineering and System Safety* 65:147-185.

Kouchmeshky, B., and N. Zabaras. 2010. Microstructure Model Reduction and Uncertainty Quantification in Multiscale Deformation Process. *Computational Materials Science* 48(2):213-227.

Lemieux, C. 2009. *Monte Carlo and Quasi-Monte Carlo Sampling*. Springer Series in Statistics. New York: Springer.

Lieberman, C., K. Wilcox, and O. Ghattas. 2010. Parameter and State Model Reduction for Large-Scale Statistical Inverse Problems. *SIAM Journal on Scientific Computing* 32(5):2523-2542.

Lin, C.D., D. Bingham, R.R. Sitter, and B. Tang. 2010. A New and Flexible Method for Constructing Designs for Computer Experiments. *Annals of Statistics* 38(3):1460-1477.

Marchuk, D.I. 1995. *Adjoint Equations and Analysis of Complex Systems*. Dordrecht, The Netherlands: Kluwer Academic Publishers.

Marzouk, Y.M., and H.N. Najm. 2009. Dimensionality Reduction and Polynomial Chaos Acceleration of Bayesian Inference in Inverse Problems. *Journal of Computational Physics* 228:1862-1902.

McKay, M.D., R.J. Beckman, and W.J. Conover. 1979. A Comparison of Three Methods for Selecting Values of Input Variables in the Analysis of Output from a Computer Code. *Technometrics* 21(2):239-245.

Mitchell, T.J., M.D. Morris, and D. Ylvisaker. 1994. Asymptotically Optimum Experimental Designs for Prediction of Deterministic Functions Given Derivative Information. *Journal of Statistical Planning and Inference* 41:377-389.

Moore, L.M. 1981. Principle Component Analysis in Linear Systems: Controllability, Observability, and Model Reduction. *IEEE Transactions on Automatic Control* 26(1):17-31.

Moore, L.M., and M.D. McKay. 2002. Orthogonal Arrays for Computer Experiments to Assess Important Inputs. D.W. Scott (Ed.). Pp. 546-551 in *Proceedings of PSAM6, 6th International Conference on Probabilistic Safety Assessment and Management*.

Morris, M. 1991. Factorial Sampling Plans for Preliminary Computational Experiments. *Technometrics* 33(2):161-174.

Najm, H.N. 2009. Uncertainty Quantification and Polynomial Chaos Techniques in Computational Fluid Dynamics. *Annual Review of Fluid Mechanics* 41:35-52.

Neal, R.M. 1993. *Probabilistic Inference Using Markov Chain Monte Carlo Methods*. CRG-TR-93-1. Toronto, Canada: Department of Computer Science, University of Toronto.

Noor, A.K., and J.M. Peters. 1980. Nonlinear Analysis via Global-Local Mixed Finite Element Approach. *International Journal for Numerical Methods in Engineering* 15(9):1363-1380.

Oakley, J. 2004. Estimating Percentiles of Uncertain Computer Code Outputs. *Journal of the Royal Statistical Society: Series C (Applied Statistics)* 53(1):83-93.

Oakley, J., and A. O'Hagan. 2002. Bayesian Inference for the Uncertainty Distribution of Computer Model Outputs. *Biometrika* 89(4):769-784.

Oakley, J., and A. O'Hagan. 2004. Probabilistic Sensitivity Analysis of Complex Models: A Bayesian Approach. *Journal of the Royal Statistical Society: Series B (Statistical Methodology)* 66(3):751-769.

Owen, A.B. 1997. Monte Carlo Variance of Scrambled Net Quadrature. *SIAM Journal on Numerical Analysis* 34(5):1884-1910.

Ranjan, P., D. Bingham, and G. Michailidis. 2008. Sequential Experiment Design for Contour Estimation from Complex Computer Codes. *Technometrics* 50(4):527-541.

Rewienski, M., and J. White. 2003. A Trajectory Piecewise-Linear Approach to Model Order Reduction and Fast Simulation of Nonlinear Circuits and Micromachined Devices. *IEEE Transactions on Computer-Aided Design of Integrated Circuits and Systems* 22(2):155-170.

Sacks, J., W.J. Welch, T.J. Mitchell, and H.P. Win. 1989. Design and Analysis of Computer Experiments. *Statistical Science* 4(4):409-423.

Saltelli, A., and I.M. Sobol. 1995. About the Use of Rank Transformation in Sensitivity Analysis of Model Output. *Reliability Engineering and System Safety* 50(3):225-239.

Saltelli, A., K. Chan, and E.M. Scott. 2000. *Sensitivity Analysis. Wiley Series in Probability and Statistics*, Vol. 535. Hoboken, N.J.: Wiley.

Seber, G.A.F., and C.J. Wild. 2003. *Nonlinear Regression*. Hoboken, N.J.: Wiley.

Shahabuddin, P. 1994. Importance Sampling for the Simluation of Highly Reliable Markovian Systems. *Management Science* 40(3):333-352.

Sirovich, L. 1987. Turbulence and the Dynamics of Coherent Structures. Part I: Coherent Structures. *Quarterly Journal of Applied Mathematics* XLV(2):561-571.

Sobol, W.T. 1993. Analysis of Variance for "Component Stripping" Decomposition of Multiexponential Curves. *Computer Methods and Programs in Biomedicine* 39(3-4):243-257.

Soize, C., and R. Ghanem. 2004. Physical Systems with Random Uncertainties: Chaos Representations with Arbitrary Probability Measure. *SIAM Journal on Scientific Computing* 26(2):395-410.

Tang, B. 1993. Orthogonal Array-Based Latin Hypercubes. *Journal of the American Statistical Association* 88(424):1392-1397.

Weickum, G., M.S. Eldred, and K. Maute. 2006. Multi-Point Extended Reduced-Order Modeling for Design Optimization and Uncertainty Analysis. Paper AIAA-2006-2145. *Proceedings of the 47th AIAA/ASME/ASCE/AHS/ASC Structures, Structural Dynamics, and Materials Conference (2nd AIAA Multidisciplinary Design Optimization Specialist Conference)*. May 1-4, 2006, Newport, R.I.

Xiu, D. 2007. Efficient Collocational Approach for Parametric Uncertainty Analysis. *Communications in Computational Physics* 2:293-309.

Xiu, D. 2010. *Numerical Methods for Stochastic Computations: A Spectral Method Approach*. Princeton, N.J.: Princeton University Press.

Xiu, D., and G.E. Karniadakis. 2002. Modeling Uncertainty in Steady State Diffusion Problems via Generalized Polynomial Chaos. *Computer Methods in Applied Mechanics and Engineering* 191(43):4927-4943.

5

Model Validation and Prediction

5.1 INTRODUCTION

From a mathematical perspective, validation is the process of assessing whether or not the quantity of interest (QOI) for a physical system is within some tolerance—determined by the intended use of the model—of the model prediction. Although "prediction" sometimes refers to situations where no data exist, in this report it refers to the output of the model in general.

In simple settings validation could be accomplished by directly comparing model results to physical measurements for the QOI and computing a confidence interval for the difference, or carrying out a hypothesis test of whether or not the difference is greater than the tolerance (see Oberkampf and Roy, 2010, Chapter 12). In other settings, a more complicated statistical modeling formulation may be required to combine simulation output, various kinds of physical observations, and expert judgment to produce a prediction with accompanying prediction uncertainty, which can then be used for the assessment. This more complicated formulation can also produce predictions for system behavior in new domains where no physical observations are available (see Bayarri et al., 2007a; Wang et al., 2009; or the case studies of this chapter).

Assessing prediction uncertainty is crucial for both validation (which involves comparison with measured data) and prediction of yet-unmeasured QOIs. This uncertainty typically comes from a number of sources, including:

- Input uncertainty—lack of knowledge about parameters and other model inputs (initial conditions, forcings, boundary values, and so on);
- Model discrepancy—the difference between model and reality (even at the best, or most correct, model input settings);
- Limited evaluations of the computational model; and
- Solution and coding errors.

In some cases, the verification effort can effectively eliminate the uncertainty due to solution and coding errors, leaving only the first three sources of uncertainty. Likewise, if the computational model runs very quickly, one could evaluate the model at any required input setting, eliminating the need to estimate what the model would have produced at an untried input setting.

The process of validation and prediction, explored in previous publications (e.g., Klein et al., 2006; NRC, 2007, Chapter 4), is described in this chapter from a more mathematical perspective. The basic process includes identifying and representing key sources of uncertainty; identifying physical observations; experiments, or other information sources for the assessment; assessing prediction uncertainty; assessing the reliability or quality of the prediction; supplying information on how to improve the assessment; and communicating results.

Identifying and representing uncertainties typically involves sensitivity analysis to determine which features or inputs of the model affect key model outputs. Once they are identified, one must determine how best to represent these important contributors to uncertainty—parametric representations of input conditions, forcings, or physical modeling schemes (e.g., turbulent mixing of fluids). In addition to parametric forms, some analyses might assess the impact of alternative physical representations/schemes within the model. If solution errors or other sources of model discrepancy are likely to be important contributors to prediction uncertainty, their impact must also be captured in some way.

The available physical observations are key to any validation assessment. In some cases these data are observational, provided by nature (e.g., meteorological measurements, supernova luminosities); in other cases, data come from a carefully planned hierarchy of controlled experiments—e.g., the Predictive Engineering and Computational Sciences (PECOS) case study in Section 5.9. In addition to physical observations, information may come from the literature or expert judgment that may incorporate historical data or known physical behavior.

Estimating prediction uncertainty requires the combination of computational models, physical observations, and possibly other information sources. Exactly how this estimation is carried out can range from very direct, as in the weather forecasting example in Figure 5.1, to quite complicated, as described in the case studies in this chapter. In these examples, some physical observations are used to refine or constrain uncertainties that contribute to prediction uncertainty. Estimating prediction uncertainty is a vibrant research topic whose methods vary depending on the features of the problem at hand.

For any prediction, assessing the quality, or reliability, of the prediction is crucial. This concept of prediction reliability is more qualitative than is prediction uncertainty. It includes verifying the assumptions on which an estimate is based, examining the available physical measurements and the features of the computational model, and applying expert judgment. For example, well-designed sets of experiments can lead to stronger statements regarding the quality and reliability of more extrapolative predictions, as compared to observational data from a single source. Here the concept of "nearness" of the physical observations to the predictions of the intended use of the model becomes relevant, as does the notion of the domain of applicability for the prediction. However,

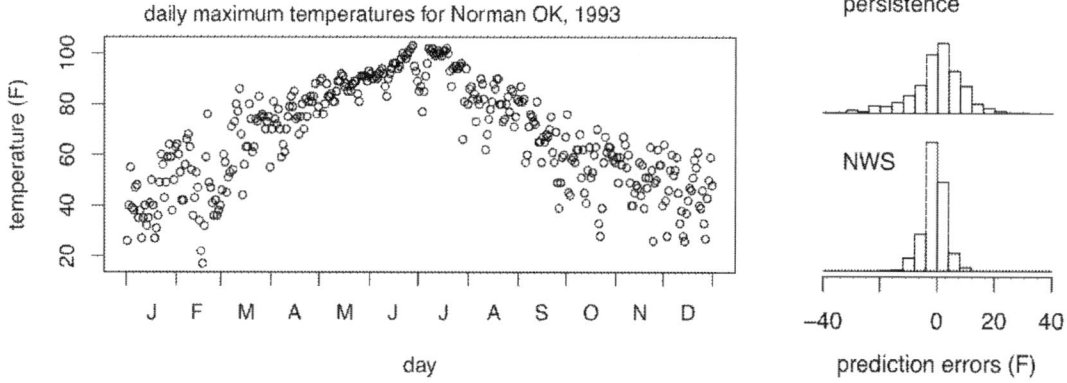

FIGURE 5.1 Daily maximum temperatures for Norman, Oklahoma (*left*), and histograms of next-day prediction errors (*right*) using two prediction models. The top histogram shows residuals from the persistence model, predicting tomorrow's high temperature with today's high temperature. The bottom histogram shows residuals from the National Weather Service (NWS) forecast. Ninety percent of the actual temperatures are within ±14°F for the persistence-model forecasts and within ±6°F for the NWS forecasts. The greater accuracy of the NWS forecasts is due to NWS's use of computational models and additional meteorological information. The assessment of these two forecast methods is relatively straightforward because of the large number of comparisons of model forecast to measurement. SOURCE: Data from Brooks and Doswell (1996).

while most practitioners recognize that this concept and notion are important, rigorous mathematical definitions and quantifications remain an unsolved problem.

In some validation applications, an opportunity exists to carry out additional experiments to improve the prediction uncertainty and/or the reliability of the prediction. Estimating how different forms of additional information would improve predictions or the validation assessment can be an important component of the validation effort, guiding decisions about where to invest resources in order to maximize the reduction of uncertainty and/or an increase in reliability.

Communicating the results of the prediction or validation assessment includes both quantitative aspects (the predicted QOI and its uncertainty) and qualitative aspects (the strength of the assumptions on which the assessment is based). While the communication component is not fundamentally mathematical, effective communication may depend on mathematical aspects of the assessment.

The various tasks mentioned in the preceding paragraphs give a broad outline of validation and prediction. Exactly how these tasks are carried out depends on features of the specific application. The list below covers a number of important considerations that will have an impact on the methods and approaches for carrying out validation and prediction:

- The amount and relevance of the available physical observations for the assessment,
- The accuracy and uncertainty accompanying the physical observations,
- The complexity of the physical system being modeled,
- The degree of extrapolation required for the prediction relative to the available physical observations and the level of empiricism encoded in the model,
- The computational demands (run time, computing infrastructure) of the computational model,
- The accuracy of the computational model's solution relative to that of the mathematical model (numerical error),
- The accuracy of the computational model's solution relative to that of the true, physical system (model discrepancy),
- The existence of model parameters that require calibration using the available physical observations, and
- The availability of alternative computational models to assess the impact of different modeling schemes or physics implementations on the prediction.

These considerations are discussed throughout this chapter, which describes key mathematical issues associated with validation and prediction, surveying approaches for constraining and estimating different sources of prediction uncertainty. Specifically, the chapter briefly describes issues regarding measurement uncertainty (Section 5.2), model calibration and parameter estimation (Section 5.3), model discrepancy (Section 5.4), and the quality of predictions (Section 5.5), focusing on their impact on prediction uncertainty. These concepts are illuminated by two simple examples (Boxes 5.1 and 5.2) that extend the ball-drop example in Chapter 1, and by two case studies (Sections 5.6 and 5.9). Leveraging multiple computational models (Section 5.7) and multiple sources of physical observations (Section 5.8) is also covered, as is the use of computational models for aid in dealing with rare, high-consequence events (Section 5.10). The chapter concludes with a discussion of promising research directions to help address open problems.

5.1.1 Note Regarding Methodology

Most of the examples and case studies presented in this chapter use Bayesian methods (Gelman et al., 1996) to incorporate the various forms of uncertainty that contribute to the prediction uncertainty. Bayesian methods require a prior description of uncertainty for the uncertain components in a formulation. The resulting estimates of uncertainty—for parameters, model discrepancy, and predictions—will depend on the physical observations and the details of the model formulation, including the prior specification. This report does not go into such details but points to references on modeling and model checking from a Bayesian perspective (Gelman et al., 1996; Gelfand and Ghosh, 1998). While the Bayesian approach is prevalent in the VVUQ literature, effectively dealing with many

Box 5.1
The Ball-Drop Experiment Using a Variety of Balls

In addition to the measurements of drop times for the bowling ball, we now have measurements for a basketball and baseball as well. The measured drop times are normally distributed about the true time, with a standard deviation of 0.1 seconds. The QOI is the drop time for the softball—an untested ball—at a height of 100 m. This QOI is an extrapolation in two ways: no drops over 60 m have been carried out; no measurements have been obtained for a softball.

The conceptual/mathematical model (Figure 5.1.1(b)) accounts for acceleration due to gravity g and air resistance using a standard model. Air resistance depends on the radius and density of the ball (R_{ball}, ρ_{ball}), as well as the density of the air (ρ_{air}). Figure 5.1.1(a) shows various balls and their position in radius-density space. It is assumed that air density is known. In addition to depending on the descriptors of the ball (R_{ball}, ρ_{ball}), the model also depends on two parameters—the acceleration of gravity g and a dimensionless friction coefficient C_D—which need to be constrained with measurements. Initial ranges of $8 \leq g \leq 12$ and $0.2 \leq C_D \leq 2.0$ are specified for the two model parameters. Measured drop times from heights of 20, 40, and 60 m are obtained for the basketball and baseball; measured drop times from heights of 10, 20, . . . , 60 m are obtained for the bowling ball. These measurements constrain the uncertainty of the parameters to the ellipsoidal region shown in Figure 5.1.1(c).

Figure 5.1.1(d) shows initial and constrained prediction uncertainties for the four different balls using the mathematical model in Figure 5.1.1(b). The light lines correspond to the parameter settings depicted by the points in Figure 5.1.1(c). The dark region shows prediction uncertainty induced by the constrained uncertainty for the parameters. A prediction (with uncertainty) for the softball is given by the spread of the dark region of the rightmost frame.

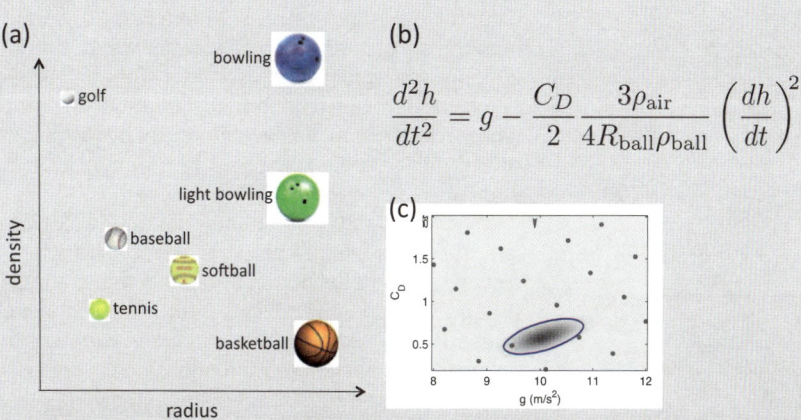

However, the model has never been tested against drops higher than 60 m. It has also never been directly compared to any softball drops. From Figure 5.1.1(a), one could argue that the softball is at the interior of the (R_{ball}, ρ_{ball})-space spanned by the basketball, baseball, and bowling ball, leading one to trust the prediction (and uncertainty) for the softball at 40 m, or even 100 m. However, the softball differs from these other balls in more ways than just radius and density (e.g., surface smoothness). How should one modify predictions and uncertainties to account for these flavors of extrapolation? This is an open question in V&V and UQ research.

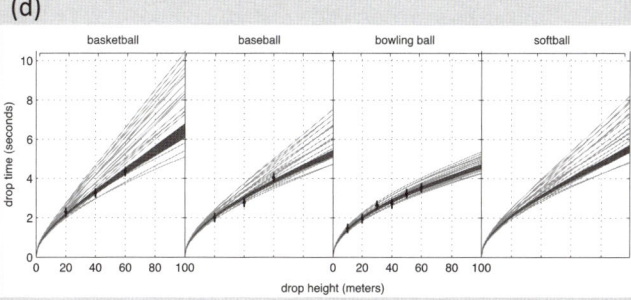

FIGURE 5.1.1

Box 5.2
Using an Emulator for Calibration and Prediction with Limited Model Runs

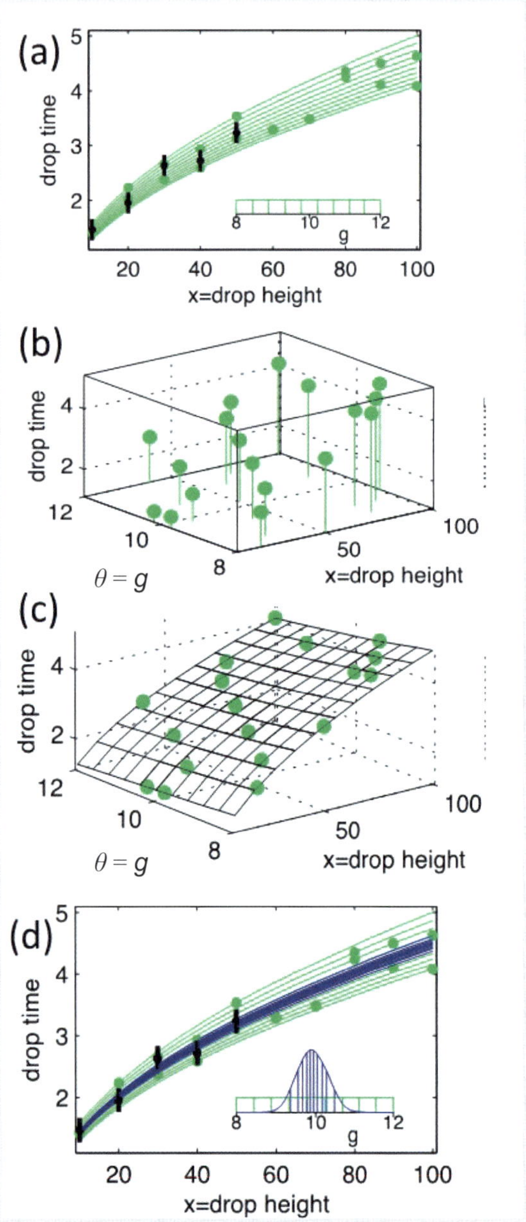

FIGURE 5.2.1

Physical measurements (black dots) and prior prediction uncertainty (green lines) for the bowling ball drop time as a function of height (as shown in Box 1.1). The experimentally measured drop times for drops of 10, 20, . . . , 50 m are shown in Figure 5.2.1(a); the uncertainty due to prior uncertainty for gravity g is also shown in the inset figure.

If the number of computer model runs were limited—perhaps due to computational constraints—then an ensemble of runs could be carried out at different (x, θ) input settings. Figure 5.2.1(b) shows model runs carried out over a statistical design of 20 input settings. Here x denotes height and θ denotes the model parameter g. The modeled drop times at these input settings are given by the height of the circle plotting symbols in Figures 5.2.1(a) and (b).

With these 20 computer model runs, a Gaussian process is used to produce a probabilistic prediction of the model output at untried input settings (x, θ), as shown in Figure 5.2.1(c). This emulator is used to facilitate the computations required to estimate the posterior distribution for θ, which is constrained by the physical observations.

The Bayesian model formulation, with an emulator to assist with limited model runs, produces a posterior distribution for the unknown parameter θ (g, given by the blue lines of the inset in Figure 5.2.1(d)), which then can be propagated through the emulator to produce constrained, posterior prediction uncertainties (blue lines).

issues discussed here, the use of these methods in the examples and case studies in this chapter should not be seen as an exclusive endorsement of Bayesian methods over other approaches for calculating with and representing uncertainty, such as likelihood (Berger and Wolpert, 1988), Dempster-Shafer theory (Shafer, 1976), possibility theory (Dubois et al., 1988), fuzzy logic (Klir and Yuan, 1995), probability bounds analysis (Ferson et al., 2003), and so on. The committee believes that relevance of the main issues discussed in this chapter is not specific to the details of how uncertainty is represented.

5.1.2 The Ball-Drop Example Revisited

To elaborate these ideas, an extension of the simple ball-drop example from Box 1.1 in Chapter 1 is used; the experiment here includes multiple types of balls (Box 5.1). Drop times for balls of various radii and densities are considered. The basic model that assumes only acceleration due to gravity is clearly insufficient when considering balls of various sizes and densities, suggesting the need for a model that explicitly accounts for drag due to air friction. This new model describes initial conditions for a single experiment, with the radius of the ball R_{ball} and the density of the ball ρ_{ball}. The model also has two parameters—the acceleration due to gravity, g, and a friction constant, C_D—that can be further constrained, or calibrated, using experimental measurements. Of course, treating the acceleration due to gravity g as something uncertain may not be appropriate in a serious application, since this quantity has been determined experimentally with very high accuracy. The motivation for treating g as uncertain is to illustrate issues regarding uncertain physical constants, which are common in many applications.

Using measured drop times for three balls—a bowling ball, baseball, and basketball—the object is to predict the drop time for a softball at 100 meters (m). Hence the QOI is the drop time for a softball dropped from a height of 100 m. Drops are conducted from a 60 m tower. The required prediction is an extrapolation in two ways: no drops over 60 m have been carried out, and no drop-time measurements have been obtained for the softball. Section 5.5 looks more closely at how validation and UQ approaches depend on the availability of measurements and the degree of extrapolation associated with the prediction.

Initially, the uncertainty about the uncertain model parameters is that $8 < g < 12$, and $0.2 < C_D < 2$, which is given by the equation in Figure 5.1.1(b). Model predictions can be made using various (g, C_D) values over this region (the dots in Figure 5.1.1(c)); the resulting drop-time predictions are given by the light lines in Figure 5.1.1(d). This uncertainty is obtained by simple forward propagation of the uncertainty in g and C_D, as described in Section 4.2 in Chapter 4. If the validation assessment were a question of whether or not the model can predict the 100 m softball drop time to within ± 2 seconds, or whether the drop time will be larger than 10 seconds, this preliminary assessment might be sufficient. If more accuracy is required, the uncertainty in the parameters (g, C_D) can be further constrained using the observed drop times for the different balls, as given by the ellipse in Figure 5.1.1(c), showing a 95 percent probability range for (g, C_D). This process of constraining parameter uncertainties using experimental measurements is called *model calibration*, or *parameter estimation*, and is discussed in more detail in Section 5.3. Physical measurements are uncertain, each giving an imperfect interrogation of the physical system, and this uncertainty affects how tightly these measurements constrain parameter uncertainty. Measurement uncertainty also plays an important role in the comparison of model prediction to reality. This topic is discussed briefly in Section 5.2.

Although the ball-drop example used here does not show any evidence of a systematic *discrepancy* between model and reality, such discrepancies are common in practice. Once identified and quantified, systematic model discrepancy can be accounted for to improve the model-based predictions (e.g., a computational-model prediction that is systematically 10 percent too low for a given QOI can simply be adjusted up by 10 percent to predict reality more accurately). Section 5.4 discusses the related idea of making the best predictions that one can with an imperfect model (and quantifying their uncertainties), embedded within a statistical framework aided by subject-matter knowledge and available measurements.

The relevant *body of knowledge* in the ball-drop example consists of measurements from three basketball drops, three baseball drops, and six bowling-ball drops, along with the mathematical and computational models. The friction term in the model is an effective physics model, slowing the ball as it drops and attempting to capture small-scale effects of airflow around the ball. Experience suggests that the friction constant depends on the velocity and smoothness of the ball, as well as on properties of the air. Ideally, part of the assessment of the uncertainty about the QOI (softball drop time from 100 m) will include at least a qualitative assessment of the appropriateness of using this form of friction model, with a single value for C_D, for these drops. This notion of assessing the reliability, or quality, of a model-based prediction is discussed in Section 5.5.

More generally, the body of knowledge could include a variety of information sources, ranging from experimental measurements to expert judgment, to results from related studies. Some of these information sources may be used explicitly, constraining parameter uncertainties, estimating variances, or describing prediction uncertainties. Other information sources might lend evidence to support assumptions used in the analysis, such as the adequacy of the model for predictions that move away from the conditions in which experimental measurements are available.

Ideally, the *domain of applicability* for this model in predicting drop times of various balls will also be specified. For example, given the current body of knowledge, a conservative domain of applicability might include only basketballs, baseballs, and bowling balls dropped from heights between 10 m and 60 m. In this case, one would not be willing to use the model-based uncertainty given in Box 5.1 to characterize the drop time for a softball at 40 m, let alone 100 m. A more liberal definition of the domain of applicability might be any ball with a radius-density combination in the interior of the basketball-baseball-bowling ball triangle in Figure 5.1.1(a).

Alternatively, one might also consider what perturbations of a basketball, say, would be included in this domain of applicability. Should predictions and uncertainties for a slightly smaller basketball be trusted? What about a slightly less dense basketball? At what density should the predictions and uncertainties no longer be trusted? Put differently, can we assess what perturbations of a basketball are sufficiently "near" to the tested basketball to result in accurate predictions and uncertainty estimates? Often, a sensitivity analysis (SA) can help address the question—this example informing trust in model-based predictions and uncertainties for balls as density decreases. One might also consider conditions that are not accounted for in the model. For example, should the drop times of a rubber basketball differ from those of a leather one? Does ball texture affect drop time? Without additional experiments, such model-applicability issues must necessarily be addressed with expert judgment or other information sources. Quantifying the impact of such issues remains an unsolved problem.

In general, the domain of applicability describes the conditions over which the predictions and uncertainties derived from a computational model are reliable. This should include descriptors of the initial conditions that are accounted for in the model, as well as those that are not. It might also include descriptors of the geometric and/or physical complexity of the system for which the prediction is being made. Such considerations are crucial for designing a series of validation experiments to help map out this domain of applicability. Defining this domain of applicability depends on the available body of knowledge, including subject-matter expertise, and involves a number of qualitative features about the inference being made.

5.1.3 Model Validation Statement

In summary, validation is a process, involving measurements, computational modeling, and subject-matter expertise, for assessing how well a model represents reality for a specified QOI and domain of applicability. Although it is often possible to demonstrate that a model does not adequately reproduce reality, the generic term *validated model* does not make sense. There is at most a body of evidence that can be presented to suggest that the model will produce results that are consistent with reality (with a given uncertainty).

Finding: A simple declaration that a model is "validated" cannot be justified. Rather, a validation statement should specify the QOIs, accuracy, and domain of applicability for which it applies.

The body of knowledge that supports the appropriateness of a given model and its ability to predict the QOI in question, as well as the key assumptions used to make the prediction, is important information to include in the reporting of model results. Such information will allow decision makers to better understand the adequacy of the model, as well as the key assumptions and data sources on which the reported prediction and uncertainty rely.

The degree to which available physical data are relevant to the prediction of interest is a key concept in the V&V literature (Easterling, 2001; Oberkampf et al., 2004; Klein et al., 2006). How one uses the available body of knowledge to help define this domain of validity is part of how the argument for trust in model-based prediction is constructed. This topic is explored further in Section 5.5.

5.2 UNCERTAINTIES IN PHYSICAL MEASUREMENTS

Throughout this chapter, reference is continually made to learning about the computational model and its uncertainties through comparing the predictions of the computational model to available physical data relevant to the QOI. A complication that typically arises is that the physical measurements are themselves subject to uncertainties and possibly bias. In the ball-drop example in Box 5.1, for instance, there were three multiple observations for each type of ball drop, and these were believed to be normally distributed, centered at the true drop time and with standard deviation of 0.1 seconds. The uncertainty in the physical measurements was part of the reason that the parameters in the example were constrained only to the ellipse in Figure 5.1.1(c) and not to a smaller area.

Although the characterization of such measurement uncertainty is often a crucial part of a VVUQ analysis, the issue is not highlighted in this report because such characterization is the standard domain of statistics, and vast methodology and experience exist for characterizing such uncertainty (Youden, 1961, 1972; Rabinovich, 1995; Box et al., 2005). However, there are several issues that must be kept in mind when obtaining physical data for use in VVUQ analyses.

For experiments that have not yet been performed, the design of the experiment for collecting the physical data should be developed in cooperation with the VVUQ analyst and the decision maker to provide maximum VVUQ benefit when practical. Experimental data are often expensive (as when each data point arises from crashing a prototype vehicle, for instance) and should be chosen to provide optimal information from the perspective of the desired calibration, VVUQ analysis, and/or the prediction for the computational model.

One particularly relevant consideration in the context of VVUQ is the desirability of replications[1] of the physical measurements—that is, of obtaining repeat measurements under the same conditions (same model-input values). This might seem counterintuitive from the perspective of the computational model; if the analyst is trying to judge how well the model predicts reality, observing reality at as many input values as possible would seem logical. When the physical data are subject to measurement error, however, the picture changes, because it is first crucial to learn how well the physical data represent reality. If the physical data do not constrain reality significantly at any input values, little has been learned that will help in judging the fidelity of the computational model with respect to reality.

If the measurement error of the physical data and variability of the physical system are known (e.g., the data has a known standard deviation) and are judged to be small enough to adequately constrain reality, then replicate observations are perhaps not needed. However, it is wise to view the presumption of known standard deviation with healthy skepticism. When the magnitude of the measurement error is derived from the properties of measurement apparatus and theoretical considerations, it is common to miss important sources of variation and bias that are present in the measurement process. Hence, resources may be better spent obtaining replicate observations, rather than attempting to account for every possible source of uncertainty present in a single measurement/experiment. One may be able to afford only enough physical data with replications to adequately constrain reality at a few input values, but knowing reality, with accurately quantified uncertainty, at a few input values is often better than having a vague idea about reality at many input values.

One does not always have control over the process of obtaining physical measurements. They may have been based on historical experiments or observations, for which important details may be unknown. They may have arisen from auxiliary inverse-problem analyses (e.g., inferring a quantity such as temperature or contaminant concentration from remotely sensed signals). This inexactness can be problematic from a number of perspectives, including the possibility that uncertainties in the physical data may have been estimated poorly, or not given at

---

[1] Here we mean *genuine replicates* as described in Box and Draper (1987, p. 71): "Replicate runs must be subject to all the usual setup errors, sampling errors, and analytical errors which affect runs made at different conditions. Failure to achieve this will typically cause underestimation of the error and will invalidate the analysis."

all. In such cases it may be fruitful to include this auxiliary inverse problem as part of the validation and prediction process.

A significant issue that can arise is possible bias in the physical data, wherein a common error induces a similar effect on all of the measurements. In the ball-drop example, for instance, a bias in the physical observations would be present if the stopwatch used to time all of the drops were systematically slow. Similarly, if each ball were released with a slight downward velocity, then measured drop times would be systematically too short.

The methodological issue of how to incorporate uncertainty in the physical data into the UQ analysis is also important. Standard statistical techniques can allow one to summarize the physical data in terms of the constraints that they place on reality, but a VVUQ analysis requires interfacing this uncertainty with the computational model, especially if calibration is also being done based on the physical data. Bayesian analysis (discussed in Section 5.3) has the appeal of providing a direct methodology for such incorporation of uncertainty.

5.3 MODEL CALIBRATION AND INVERSE PROBLEMS

Many applications in VVUQ use physical measurements to constrain uncertain parameters in the computational model. A simple example is given in Figure 5.1.1(c), in which measured drop times are used to reduce the uncertainty in the two model parameters—g and C_D. This basic task of model calibration is a standard problem in statistical inference. Model calibration applications may involve parameters ranging from one or two, as in Box 5.1, to thousands or millions, as is often the case when one is inferring heterogeneous fields (material properties, initial conditions, or source terms—e.g., Akçelik et al., 2005).

The problem of estimating from observations the uncertain parameters in a simulation model is fundamentally an inverse problem. The *forward problem* seeks to predict output observables (such as seismic ground motion at seismometer locations) given the parameters (such as the heterogeneous elastic-wave speeds and density throughout a region of interest) by solving the governing equations (such as the elastic-wave equations). The forward problem is usually well posed (the solution exists, is unique, and is stable to perturbations in inputs), causal (later-time solutions depend only on earlier-time solutions), and local (the forward operator includes derivatives that couple nearby solutions in space and time).

The *inverse problem* reverses this relationship, however, by seeking to determine parameter values that are consistent with particular measurements. Solving inverse problems can be very challenging for the following reasons: (1) the mapping from observations (i.e., measurements) to parameters may not be one to one, particularly when the number of parameters is large and the number of measurements is small; (2) small changes in the measurement value may lead to changes in many or all parameters, particularly when the forward model is nonlinear; and (3) typically, all that is available to the analyst is a computational model that approximately solves the forward problem.

In simple model calibration, or inverse problems, post-calibration parameter uncertainty can be described by a "best estimate" of uncertainty determined by a covariance matrix, characterizing variance and correlations in the parameter uncertainties. When the solution to the inverse problem is not unique, and/or when the measurement errors have a nonstandard form, determining even a best estimate can be problematic. The popular approach to obtaining a unique "solution" to the inverse problem in these circumstances is to formulate it as an optimization problem—minimize the sum of two terms: the first is a combination of the misfit between observed and predicted outputs in an appropriate norm, and the second is a regularization term that penalizes unwanted features of the parameters. This is often called Occam's approach—find the "simplest" set of parameters that is consistent with the measured data. The inverse problem thus leads to a nonlinear optimization problem in which the forward simulation model is embedded in the misfit term. When the forward model takes the form of partial differential equations (PDEs) or some other expensive model, the result is an optimization problem that may be extremely large scale in the state variables (displacements, temperatures, pressure, and so on), even when the number of inversion parameters is small. More generally, uncertain parameters can be taken from numbers on a continuum (such as initial or boundary conditions, heterogeneous material parameters, or heterogeneous sources) that, when discretized, result in an inverse problem that is very large scale in the inversion parameters as well.

An estimation of parameters using the regularization approach to inverse problems as described above will yield an estimate of the "best" parameter values that minimize the combined misfit and penalty function.

However, in UQ, the analyst is interested not just in point estimates of the best-fit parameters but also in a complete statistical description of all parameter values that are consistent with the data. The Bayesian approach does this by reformulating the inverse problem as a problem in statistical inference, incorporating uncertainties in the measurements, the forward model, and any prior information about the parameters. The solution of this inverse problem is the set of so-called posterior probability densities of the parameters, describing updated uncertainty in the model parameters (Kaipio and Somersalo, 2005; Tarantola, 2005). Thus the resulting uncertainty in the model parameters can be quantified, taking into account uncertainties in the data, uncertainties in the model, and prior information. The term *parameter* is used here in the broadest sense and includes initial and boundary conditions, sources, material properties and other coefficients of the model, and so on; indeed, Bayesian methods have been developed to infer uncertainties in the form of the model as well (so-called structural uncertainties or model inadequacy are discussed in Section 5.4).

The Bayesian solution of the inverse problem proceeds as follows. Let the relationship between model predictions of observable outputs y and uncertain input parameters θ be denoted by

$$y = f(\theta, e)$$

where e represents noise due to measurement and/or modeling errors. In other words, given the parameters θ, the function $f(\theta)$ invokes the solution of the forward problem to yield y, the predictions of the observables. Suppose that the analyst has a prior probability density $\pi_{pr}(\theta)$, which encodes the prior information about the unknown parameters (i.e., independent of information from the present observations). Suppose further that the analyst can build—using the computational model—the likelihood function $\pi(y_{obs}|\theta)$, which describes the conditional probability that the parameters θ gave rise to the actual measurements y_{obs}. Then Bayes's theorem expresses the posterior probability density of the parameters, π_{post}, given the data y_{obs}, as the conditional probability

$$\pi_{post}(\theta) := \pi(\theta|y) \propto \pi_{pr}(\theta)\pi(y_{obs}|\theta) \tag{5.1}$$

The expression (5.1) provides the statistical solution of the inverse problem as a probability density for the model parameters θ.

Although it is easy to write down expressions for the posterior probability density such as expression 5.1, making use of these expressions poses a challenge owing to the high dimensionality of posterior probability density (which is a surface of dimension equal to the number of parameters), and because the solution of the forward problem is required at each point on this surface. Straightforward grid-based sampling is out of the question for anything other than a few parameters and inexpensive forward simulations. Special sampling techniques, such as Markov chain Monte Carlo (MCMC) methods, have been developed to generate sample ensembles that typically require many fewer points than are required for grid-based sampling (Kaipio and Somersalo, 2005; Tarantola, 2005). Even so, MCMC approaches become intractable as the complexity of the forward simulations and the dimension of the parameter spaces increase. The combination of a high-dimensional parameter space and a forward model that takes hours to solve makes standard MCMC approaches computationally infeasible.

As discussed in Chapter 4, one of the keys to overcoming this computational bottleneck lies in examining the details of the forward model and effectively exploiting its structure in order to reduce implicitly or explicitly the dimension of both the parameter space and the state space. The motivation for doing so is that the data are often informative about just a fraction of the "modes" of the parameter field, because the inverse problem is ill-posed. Another way of saying this is that the Jacobian of the parameter-to-observable map is typically a compact operator and thus can be represented effectively using a low-rank approximation—that is, it is often sparse with respect to some basis (Flath et al., 2011). The remaining dimensions of parameter space, which cannot be inferred from the data, are typically informed by the prior; however, the prior does not require the solution of expensive forward problems and is thus usually much cheaper to compute. Compactness of the parameter-to-observable map suggests that the state space of the forward problem can be reduced as well. Note that although generic, regularizing priors (e.g., Besag et al., 1995; Kaipio et al., 2000; Oliver et al., 1997) make posterior exploration possible, giving useful point estimates, they may not adequately describe the uncertainty in the actual field. This is common when the physical field exhibits roughness or discontinuities that are not allowed under the prior model used in the analysis. In such cases, the uncertainties produced from such an analysis will not be appropriate at small spatial scales. Such difficulties can be overcome by specifying more realistic priors.

A number of current approaches to model reduction for inverse problems show promise. These range from Gaussian process (GP) response-surface approximation of the parameter-to-observable map (Kennedy and O'Hagen, 2001); to projection-type forward-model reductions (Galbally et al., 2010; Lieberman et al., 2010); to polynomial chaos (PC) approximations of the stochastic forward problem (Badri Narayanan and Zabaras, 2004; Ghanem and Doostan, 2006; Marzouk and Najm, 2009); to low-rank approximation of the Hessian of the log-posterior (Flath et al., 2011; Martin et al., in preparation[2]). Approaches that exploit multiple model resolutions have also proven effective for speeding up MCMC in the presence of a computationally demanding forward model (Efendiev et al., 2009; Christen and Fox, 2005).

An alternative to using the standard MCMC methods on the computer model directly is to use an emulator (see Section 4.1.1, Computer Model Emulation) in its place. In many cases, this approach alleviates the computational bottleneck caused by solving the inverse problem by applying MCMC to the computer model directly. Box 5.2 shows how an emulator can reduce the number of computer model runs for the bowling ball drop application in Box 5.1.

Here the measured drop times are governed by the unknown parameters, q (the acceleration due to gravity g, for this example), and also quantities, x, that can be measured or adjusted in the physical system. For this example x denotes drop height, but more generally x might describe system geometry, initial conditions, or boundary conditions. The relationship between observable outputs and uncertain input parameters q, at a particular x, is now denoted by

$$y_{obs} = \eta(x, q) + e \tag{5.2}$$

where e denotes the measurement error. The computer model is exercised at a limited number of input configurations (x, θ), shown by the dots in Figures 5.2.1(a), (b), and (c). Next, an emulator of the computational model can be constructed and used in place of the simulator (Figure 5.2.1(b)). Alternately, the construction of the emulator and estimation of θ can be done jointly using a hierarchical model that specifies, say, a GP model for $\eta(\ )$ and treats the estimation of θ as a missing-data problem. Inferences about the parameter θ, for example, can be made using its posterior probability distribution, usually sampled by means of MCMC (Higdon et al., 2005; Bayarri et al., 2007a).

The physical observations and the computational model can be combined to estimate the parameter θ, thereby constraining the predictions of the computational model. Looking again at Figure 5.2.1(c), the probability density function (PDF) (shown by the solid curve in the center) shows the updated uncertainty for θ after combining the computational model with the physical observations. Clearly, the physical observations have greatly improved the knowledge of the unknown parameter, reducing the prediction uncertainty in the drop time for a bowling-ball drop of 100 m.

Finding: Bayesian methods can be used to estimate parameters and provide companion measures of uncertainty in a broad spectrum of model calibration and inverse problems. Methodological challenges remain in settings that include high-dimensional parameter spaces, expensive forward models, highly nonlinear or even discontinuous forward models, and high-dimensional observables, or in which small probabilities need to be estimated.

Recommendation: Researchers should understand both VVUQ methods and computational modeling to more effectively exploit synergies at their interface. Educational programs, including research programs with graduate-education components, should be designed to foster this understanding.

---

[2] Martin, J., L.C. Wilcox, C. Burstedde, and O. Ghattas, A Stochastic Newton MCMC Method for Large Scale Statistical Inverse Problems with Application to Seismic Inversion. *SIAM Journal on Scientific Computing*, to appear.

5.4 MODEL DISCREPANCY

Computer models of processes are rarely perfect representations of the real process being modeled; there is usually some discrepancy between model and reality. Although few would disagree with this near tautology, reactions are bounded between two positions: (1) "If it is the best model we can construct at the current time and with the current resources, we cannot do better than to simply use the model as a surrogate for reality" and (2) "Use of a model is never justifiable unless it has been 'proven' to be an accurate representation of reality."

The first position may appear attractive in certain applications in which one must do something (e.g., decide whether or not to evacuate because of a potential tsunami), but it often leaves much to be desired from a scientific standpoint. The second position is harder to criticize because it seems to have the ring of scientific veracity, but it can result in doing nothing when something should be done. For instance, it is unlikely that, in the near future, any climate model will be *proven* to be an accurate representation of reality, in a detailed absolute sense, yet the consequences of ignoring what the climate models suggest could be significant.

Dealing with model inadequacy is the most difficult part of VVUQ. Although one may be able to list the possible sources of model inadequacy, understanding their impact on the model's predictions of QOIs is exceedingly difficult. Furthermore, dealing with model inadequacy is arguably the most important part of VVUQ: if the model is grossly wrong because of limited capability in incorporating the physics, chemistry, biology, or mathematics, the fact that the other uncertainties in the analysis have been accounted for may be meaningless.

Formal approaches to dealing with model inadequacy can be characterized as being in one of two camps, depending on the information available. In one camp, evaluation is performed by comparing model output to physical data from the real process being modeled. The common rationale for this philosophy is that the only way to see if a model actually works is to see if its predictions are correct. This report refers to this approach as the *predictive* approach to evaluation. The other camp focuses on the model itself and tries to assess the accuracy or uncertainty corresponding to each constructed element of the model. The common rationale for this philosophy is that if the model contains all of the elements of the system it represents, if all of these elements (including computational elements) can be shown to be correct, and if they are correctly coupled, then logically the model must give accurate predictions. This report refers to this as the *logical* approach to model evaluation. Of course, any evaluation of a given model's adequacy could involve elements from each camp.

Before discussing these formal approaches, it is important to consider the metric and tolerance by which model adequacy should be measured. The obvious metric is simply accuracy in the prediction of the desired feature of the real process—the quantity of interest. The point is that no model is likely to predict every aspect of a real process accurately, but a model may accurately predict a key feature of interest of the real process, to within an acceptable tolerance for the intended application. Furthermore, because uncertainties abound, prediction inevitably has an uncertainty range attached. Thus the prediction may be in the form of a statement such as this: "The real QOI will be 5 ± 2 (with probability 0.9)." There are many advantages of viewing model adequacy through such statements. They include the following:

- Models rarely give highly accurate predictions over the entire range of inputs of possible interest, and it is difficult to characterize regions of accuracy and inaccuracy in advance. The type of statement given above indicates the accuracy of the predicted quantity of interest, and the user can decide whether the accuracy is sufficient or not.
- The degree of accuracy in the prediction (both the 2 and the 0.9 in the statement above) will typically vary from one application of the computer model to another and from one QOI to another. The degree of accuracy that is required may also differ for different intended applications and different QOIs.
- The uncertainty statements can simultaneously incorporate probabilistic uncertainty and structural uncertainty (also known as model bias or discrepancy).

Note, in particular, that the blanket statements "The model is valid" (i.e., always valid) or "The model is invalid" (i.e., always invalid) are almost always devoid of useful information (although in some cases the latter may not be). Keeping this metric of model validity in mind, there is then a formal approach to modeling inadequacy when data from the real process are available (Kennedy and O'Hagan, 2001; Higdon et al., 2005; Bayarri et al., 2007a). The automobile-suspension case study described below makes use of this approach, which is now briefly outlined.

The failure to model the system perfectly, even given the correct inputs, is due to model discrepancy, which often varies with the experimental conditions x. Let us denote the relationship between observable outputs, y, and the parameters governing the system, (x, θ), by

$$y = f(x, \theta, e)$$

where e represents noise. One formal approach for combining data from the real process with computer model runs (Kennedy and O'Hagan, 2001; Higdon et al., 2005; Bayarri et al., 2007a) views the physical observations as the sum of the computer-model output, a model inadequacy function, and noise. Mathematically this is stated as

$$y_{obs}(x) = f(x, \theta, e) = \eta(x, \theta) + \delta(x) + e$$

where $\eta(x, \theta)$ is the computer-model output with inputs (x, θ), and $\delta(x)$ is the discrepancy between the computer-model output and the true, physical mean QOI at a particular value of x (the observable or adjustable system variables).

The aim is to use realizations from the computer model and the physical observations to (1) solve the inverse problem, thereby estimating θ, (2) assess the model adequacy; and (3) build a predictive model for the system. These goals are most frequently achieved using a Bayesian hierarchical model that specifies, for example, Gaussain process (GP) models for $\eta(\ )$ and $\delta(\ )$ and treats the estimation of θ as a missing-data problem. Prior distributions must be specified for the GP parameters and also for θ. Sampling from the joint posterior distribution of these parameters is typically carried out using an MCMC algorithm, leading to estimates of all unknowns, including the discrepancy $\delta(x)$, together with error bands quantifying the uncertainties in the estimates. For alternative but related formulations for combining computational models with physical measurements for calibration and prediction, see Fuentes and Raftery (2004), Goldstein and Rougier (2004), and Tonkin and Doherty (2009).

To make these extensions more concrete, consider the reduced version of the ball-drop experiments in Box 5.3. In this illustration, the computational model is the simple model outlined in Box 5.1, and the unknown parameter in the model, θ, is the acceleration due to gravity g. This simple model fails to take into account the ball properties and the effect of air friction. Suppose for the moment that the experiments are performed only with the bowling ball (i.e., see Box 1.1 in Chapter 1, or Box 5.2) and the inverse problem is solved as outlined in Section 5.3, Model Calibration and Inverse Problems. Notice that the constrained prediction of the bowling-ball drop times performs quite well. However, the first two plots in Figure 5.3.1(a) show that the model does not accurately predict the drop times for the basketball and baseball. This would also indicate that predictions for the untested softball are problematic.

If all of the observations for the basketball, baseball, and bowling ball are available, then a model discrepancy (i.e., model inadequacy) term that accounts for each ball's radius and density can be estimated. The particular discrepancy is estimated in Figure 5.3.1(c). A glance at Figure 5.3.1(d) shows that the discrepancy-adjusted model does a much better job of predicting the physical responses. Also notice that the 95 percent probability interval for each ball is wider than those illustrated in Box 5.1. This is because the statistical model is estimating g as well as parameters for the discrepancy model (the parameters that describe the function $\alpha(R_{ball}, \rho_{ball})$).

Continuing with this illustration, suppose that interest also lies in predicting the drop times for the softball, but there are no observations available. Since the discrepancy is modeled as a function of the ball's radius and density, then discrepancy-adjusted predictions can be made for the softball (see Figure 5.3.1(d)). The 95 percent probability intervals are wider than the intervals for the other balls because there are no observations of the softball, and the inadequacy for this setting is informed through the estimated discrepancy model.

One might be tempted to use this approach to estimate drop times for the golf ball. Notice that the golf ball is not in the interior of the radius and density ranges explored in the experiments. Such a scenario constitutes an extrapolation of the discrepancy function where one is not likely to know its functional form. In any case, extrapolating beyond the experimental region should be done with great caution.

Box 5.3
Using a Model Discrepancy Term to Predict Drop Times

After the calibration of the model, which accounts only for acceleration due to gravity (Box 5.2), we find that the model does not accurately predict drop times for the basketball and baseball (Figure 5.3.1(a)). Thus the model is not considered to be adequate for predicting the drop time for a softball at 40 m or 100 m.

The conceptual and mathematical model accounts for acceleration due to gravity g only. A discrepancy-adjusted prediction is produced by adjusting the simulated drop times according to the equation:

Drop time = simulated drop time + α × drop height,

where α depends on the radius and density of the ball (R_{ball}, ρ_{ball}). The model produces an estimate for α that increases as ball density decreases (Figure 5.3.1(c)).

Figure 5.3.1(d) shows predictions using the discrepancy-adjusted model described above. The added uncertainty is due to uncertainty in both g and α in this model.

The resulting predictions and uncertainty use a model that more accurately fits known data but does not accurately reproduce reality in general. Furthermore, the discrepancy term is not physically derived. For example, this discrepancy-adjusted model does not produce a constant terminal velocity for an object that falls for a long time. This suggests that the quality of these predictions is less than those produced from the drag model used in Box 5.1. In particular, predictions for drops from greater heights, resulting in greater velocities, are suspect for this discrepancy-adjusted model. A key question is at what height (for a softball, say) this prediction (with uncertainty) becomes unreliable.

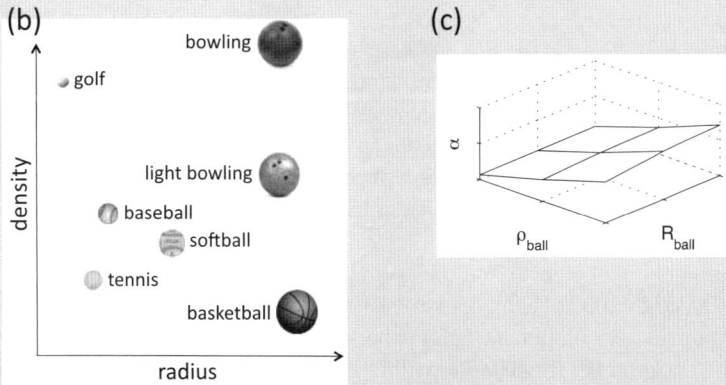

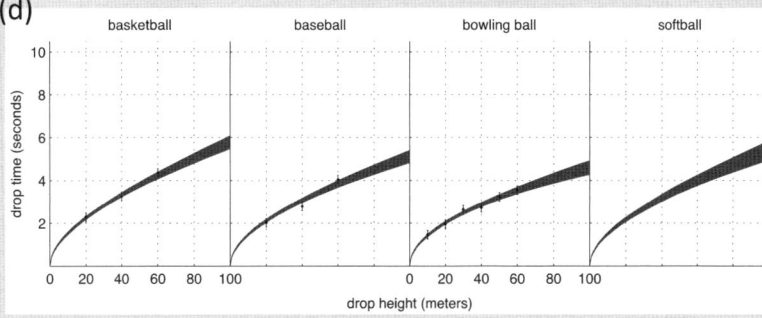

FIGURE 5.3.1

The additive discrepancy term described in this chapter—which in many published applications has been modeled as a Gaussian process—could be replaced by other, more physically motivated forms. It is also common to embed discrepancy terms within the computational-model specification. For example, the air friction term in the ordinary differential equation (ODE) in Figure 5.1.1(b) could be considered a physically motivated, embedded discrepancy term. Such terms are more commonly called parameterized, or effective, physics terms. Generally, the more physically motivated a discrepancy term is, the broader the prediction's domain of applicability. This broader perspective of constructing general approaches for adjusting basic, inadequate models to give useful predictions is likely to be a productive field for future VVUQ research.

The approach of embedding discrepancy terms within the computational-model specification is not a panacea, however, and thus it comes with a few notes of caution. For example, the approach can suffer from serious confounding between the estimates of the parameters, θ, and the estimated discrepancy function. Broadly speaking, the discrepancy can account for both good and bad choices of θ (Loeppky et al., 2011). Indeed, in cases where model inadequacy exists (i.e., the discrepancy cannot be safely estimated as zero), one should not view the estimates of the unknown parameters, θ, as solutions to the inverse problem but instead should view them more as tuning parameters to allow the sum of the computer model and discrepancy to best match the observations. Furthermore, the estimated discrepancy does not necessarily extrapolate well to new situations (although it can if one is lucky), because the "model bias or discrepancy correction" can claim to be accurate only in regions close to where there are real-process data. In spite of these issues, formally exploring the discrepancy surface can be of considerable value, exposing the regions of the input space for which the computational model is inadequate, potentially leading to opportunities for improvement in the model.

Finding: A discrepancy function can help adjust the computational model to give better interpolative predictions. A discrepancy function can also be beneficial in reducing the overtuning of parameters used to adjust or calibrate the model that can otherwise result.

If no data on the real process are available (e.g., as in the PECOS case study of Section 5.9, or for parts of the problem of determining the condition of the nuclear stockpile), there is no alternative to the "logical" approach to assessing model adequacy (although the adequacy of subcomponents of the model could be partly assessed through real-process data relevant to the subcomponents). Care is needed in analyzing the uncertainties from each possible source. It is tempting to do "worst-case" analyses, but these will often not result in useful policy guidelines because the "worst cases" are too extreme, especially if the system has a large number of components and the "worst cases" of each are combined.

For example, suppose we have a 15-component system and the probability of failure of each component is known quite accurately to be between 0.002 and 0.007. Then a worst-case analysis shows only that the probability of failure of the system (assuming it fails if any component fails) is between 0.03 and 0.10, which may be an overly wide interval for decision purposes. In contrast, if one were willing to assume that the failure probabilities of the components were (independently) uniformly distributed between 0.002 and 0.007, the resulting 95 percent confidence interval for system failure would be the interval from 0.060 to 0.070, a much smaller interval. Although one could certainly argue against the assumptions made in the latter analysis, from the decision standpoint, it might be preferable to accept such assumptions in lieu of the large uncertainties produced by the worst-case analysis.

Research and educational issues abound in the areas of accounting for model discrepancy. Below, a few main issues are summarized.

- The predictive approach to evaluating model adequacy has implementation problems: it leads to identifiability issues because of trade-offs between the model parameters and the discrepancy term, and it is difficult to implement in high-dimensional problems.
- The use of physically motivated forms for the discrepancy term, using insight from the application and from an understanding of the model's shortcomings, is an open problem that can be advanced by researchers who are versed in VVUQ methods, computational modeling, and the application at hand.

- One motivation for using multimodel ensembles (Section 5.7) is to estimate physical reality using a collection of models, each with its own discrepancy. Can ensembles of models help in estimating model discrepancy? Can one construct an ensemble of models, perhaps using bounding ideas, that allows one to quantify the difference between model predictions and reality?
- Typically, VVUQ will ultimately be tied into decision making. If model discrepancy is an important contributor to prediction uncertainty, this will need to be reflected in decision-making problems that use the model. The VVUQ analysis must produce results that can be incorporated into the encompassing decision problem.
- Assessing model adequacy should be viewed as an evolutionary process with the accumulation of evidence enhancing or degrading confidence in the model outputs and their use for an intended application. Therefore, the formal methodology must accommodate the updating of current conclusions as more information arrives.

As an alternative to the Bayesian approach, one might formally perform a hypothesis test to decide if it can be assumed that the discrepancy is zero in the tested regimes (Hills and Trucano, 2002). Two problems can arise with this approach, however. The first is that the test may suffer from a lack of statistical power. For instance, if one has physical data that are very uncertain and provide almost no constraint on reality, then the null hypothesis of zero discrepancy will not be statistically rejected, even if the computational model is extremely biased. At the opposite extreme, one might have a quite good computational model with discrepancy that is close enough to zero to make the model very useful, and yet have so much physical data that one would resoundingly reject the null hypothesis of zero discrepancy with any formal statistical test. (For an example of the latter, see Bayarri et al., 2009a.) Accordingly, the committee believes that the approach of estimating the discrepancy, together with associated error bands and specified tolerance, is the more fruitful approach.

5.5 ASSESSING THE QUALITY OF PREDICTIONS

Predictions with uncertainty are necessary for decision makers to assess risks and take actions to mitigate potential adverse events with limited resources. In addition to providing an estimate of the uncertainty, it is also crucial to assess the quality of the prediction (and accompanying uncertainty), describing and assessing the appropriateness of key assumptions on which the estimates are based, as well as the ability of the modeling process to make such a prediction. The way that one assesses the quality, or reliability, of a prediction and describes its uncertainty depends on a variety of factors, including the availability of relevant physical measurements, the complexity of the system being modeled, and the ability of the computational model to reproduce the important features of the physical system on which the QOI depends.

This section surveys issues related to assessing the quality of predictions, their prediction uncertainty, and their dependence on features of the application—including the physical measurements, the computational model, and the degree of extrapolation required to make inferences about the QOI.

For repeatable events, a computational model's predictive uncertainty can often be reliably assessed empirically, without a detailed understanding of how the model works and without a detailed understanding of how the model differs from reality. For example, consider two models for predicting tomorrow's high temperature (Figure 5.1)—one model uses today's high temperature as the prediction, and the other is the prediction provided by the National Weather Service (NWS) using state-of-the-art computational models and data feeds from ground stations and satellites. By comparing predictions to physical observations over the past year, one can infer that although both models are unbiased, the prediction from the NWS model is more accurate. A 90 percent prediction interval for the NWS predictions is ±6°F, whereas the same prediction interval for the empirical persistence model is ±14°F.

This combination of computational model with physical observations is a classic example of data assimilation. This is a mature field, with substantial literature and research focused on such filtering, or data-assimilation problems (Evensen, 2009; Welch and Bishop, 1995; Wan and Van Der Merwe, 2000; Lorenc, 2003; Naevdal et al., 2005), in which the model is repeatedly updated given new physical observations. In these applications the prediction and its uncertainty are reliably estimated, the data are relatively plentiful, and data are directly comparable to model output.

In many model-based prediction problems, a purely empirical, or statistical, approach is not feasible because there are insufficient measurement data in the domain of interest with which to assess the computational model's prediction accuracy directly. For example, consider the ball-drop experiment described in Boxes 5.1 and 5.2. The prediction as to how long it takes a softball to fall 100 m is based on a computational model and experiments involving various balls—none of which are softballs—and various drop heights, none of which are above 60 m. The automobile suspension system case study discussed in Section 5.6 is another example, as is the thermal problem described in Hills et al. (2008). In each of these examples there are limited data available that can be combined with a computational model to produce predictions and accompanying uncertainty estimates. This means that there is some degree of extrapolation involved in these predictions. Assessing the quality of the prediction and uncertainty estimate in these cases requires an understanding of the physical process and the computational model in addition to the VVUQ methodology being used.

The combination of measurement data and computational model(s) can be more intricate, as described in the PECOS Center's application on assessing the thermal protection layer of a reentry vehicle (Section 5.9), or in the cruise-missile assessment described in Oberkampf and Trucano (2000), or in the validation work of the Department of Energy (DOE) National Nuclear Security Administration (NNSA) laboratories' stockpile stewardship program (Thornton, 2011). In these applications, a hierarchy of different experiments explores different features of the physical system. Some experiments probe a single phenomenon, such as material strength or equation of state, whereas others produce measurements from processes that involve multiple physical phenomena, requiring multiphysics models. Typically, experimental measurements are more readily available for simpler experiments, involving a single effect. Highly integrated experiments are more expensive and less common. In such applications, a model to predict the QOI typically requires a multiphysics code, and the QOI is often difficult, or even impossible, to observe directly in experiments. One needs to combine measurements from these various experiments with the computational model to produce predictions with uncertainty for the QOI. Ideally, one should also assess the quality of resulting predictions and of their estimated uncertainties.

The above applications have limited data in the domain of interest, and others are even more extrapolative. The climate-modeling case study discussed in Section 2.10 of Chapter 2 is a good example of extrapolation. The QOI is global mean temperature after 15 years of doubled CO_2 forcing. In addition to extrapolations in time and forcing conditions, the predictions are based on models that do not contain all of the physical processes present in the actual climate system. The many issues in such an investigation are detailed in that case study. Other examples, such as assessing the risk of groundwater contamination from transport over the span of hundreds or thousands of years, can also be highly extrapolative. A danger in such applications is that the model may be missing key physical phenomena that are not important to the processes controlling the calibration and/or validation phase of the assessment but that are important in the system for the extrapolative prediction. Although it is quite difficult to account for potential missing processes and to quantify their effects on predicted QOIs, their existence is likely to push the model predictions away from reality in highly extrapolative settings. See Kersting et al. (1999) for a notable example in subsurface contaminant transport.

Although there does not appear to be any common, agreed-upon mathematical framework to assess the quality of validation and UQ in extrapolative situations, nearly all such applications invoke some notion of a *domain space*, describing key features of the physical and modeling processes relevant to the QOI. A very simple example is given in Box 5.2, where each experiment is described by its initial conditions—drop height, ball radius, and ball density.

This notion of the domain space is also present in the hierarchical validation scenario described in Figure 5.7, in Section 5.9.5, with the space accounting for different basic processes in the system, as well as the integration of these different processes. Figure 5.2 shows a number of domain space concepts from the VVUQ literature, ranging from specific descriptions of initial conditions to vaguer descriptors of system complexity.

This notion of domain space enables one to estimate prediction uncertainty, or quality, as a function of position in this space. In Box 5.3, the domain space, describing initial conditions, is used as the support on which the model discrepancy term is defined, enabling a quantitative description of prediction uncertainty as a function of drop height, ball radius, and ball density. Clearly, this domain space can incorporate more than just

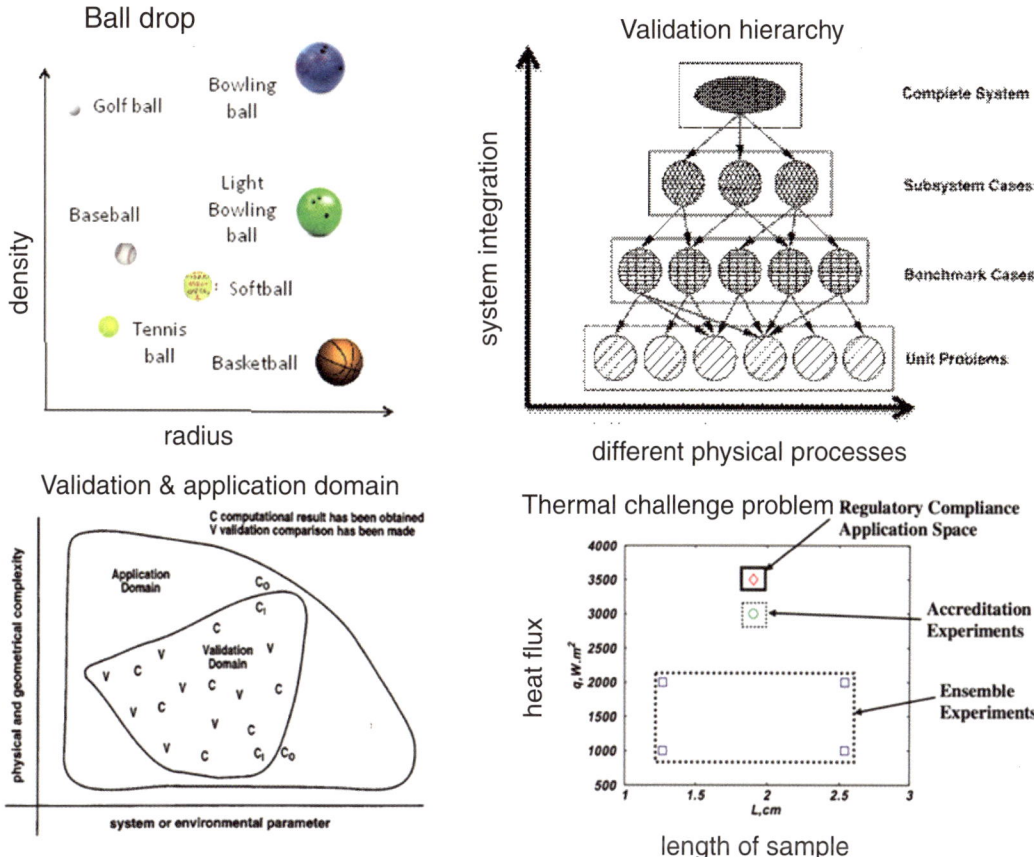

FIGURE 5.2 Domain spaces from four different VVUQ sources. The domain space describes the conditions relevant to assessing the accuracy with which a model reproduces an experiment. In some cases this domain space is very specific, describing initial conditions; in others this space is more generically specified. SOURCE: The validation hierarchy is taken from AIAA (1998); the validation and application domains are from Oberkampf et al. (2004); the thermal challenge example is from Hills et al. (2008).

model-input terms. For example, Higdon et al. (2008) uses a discrepancy term defined over two-dimensional features present in the experimental observations that cannot be incorporated into the one-dimensional model that was employed.

As Figure 5.2 suggests, it may also be fruitful to define a domain space that describes the important physical phenomena/regimes that control or affect the true physical QOI. Examples might include quantitative dimensions such as temperature and pressure visited by the physical system, as well as qualitative ones such as whether or not the QOI is affected by a phase change, turbulence, or boundary effects. By constructing such a domain space, more direct diagnoses of model shortcomings could be made. For example, if the system undergoes a phase change, then the computed QOI is not reliable since the model does not address this phenomenon. Such a phenomenon-based description of a domain space may be difficult to obtain since such features are often difficult to access experimentally. Models may help, but they may not faithfully reproduce these features of the physical system.

Mapping out such a domain space can help build understanding regarding the situations for which a computational model is expected to give sufficiently accurate predictions. It also may facilitate judgments of the nearness of available physical observations to conditions for which a model-based prediction is required. For example, subject-matter and modeling experts might agree that the model will still give reliable prediction results as certain dimensions of this space are varied, whereas with changes in other dimensions it will not. Sensitivity analysis is likely involved in specifying this domain, but it must go beyond simply exploring the model. Understanding the strengths and weakness of both the mathematical and the computational models, as they compare to reality for this application, is key. This understanding must rely heavily on subject-matter expertise.

5.6 AUTOMOBILE SUSPENSION SYSTEMS CASE STUDY

5.6.1 Background

The use of computer models of processes has enormous potential in industry for replacing costly prototype design and experimentation with much less costly computational simulations of processes. In the automotive industry, for instance, each prototype vehicle can cost hundreds of thousands of dollars to construct, and the physical testing of the vehicles is expensive. Great savings can be achieved if computer models of the vehicles, or components thereof, are used instead of prototype vehicles for design and testing. Of course, a computer model can be trusted for this only if it can be shown to provide a successful representation of the real process.

This section discusses a study that was made of a computer model of an automotive suspension system (Bayarri et al., 2007b). Of primary interest was the ability of the computer model to predict loads resulting from events stressful to the suspension system—for example, hitting a pothole. The case study provides an illustration of much of the range of needed inference in uncertainty quantification, covering the following:

- Uncertainty in model inputs,
- The need for calibration or tuning of model parameters,
- Assessment of the discrepancy between the model and the real process,
- Provision of uncertainty bounds for predictions of the model, and
- The allowing of model prediction improvements through a discrepancy adjustment.

The approach taken in the study was based on Bayesian probability analysis, which has the singular feature of allowing all of the above issues to be dealt with simultaneously and which also provided final uncertainty bounds on model predictions that account for all of the uncertainties in the inputs and model. In particular, model predictions were always presented with 90 percent confidence bands, allowing direct and intuitive assessment of whether the model predictions are accurate enough for the intended use. However, commercial software was used in the study, and so verification was not carried out, since it was assumed to be the responsibility of the software developer.

5.6.2 The Computer Model

An ADAMS[3] computer model (a commercially available, widely used finite-element-based code that analyzes the dynamic behavior of mechanical assemblies) was implemented (Bayarri et al., 2007b) to re-create the loads resulting from stresses on a vehicular suspension system.

In addition to the finite-element model itself (which must be constructed for each vehicle type), the computer model has several inputs:

- Two calibration parameters, u_1 and u_2, which quantify two types of damping (energy dissipation) that need to be estimated for (or tuned to) the physical process under study; and
- Seven unmeasured parameters of the system corresponding to characteristics of parts of the suspension system (tires, bushings, and bumpers) as well as vehicle mass; these have known nominal values but are subject to manufacturing variations and hence are treated as randomly varying around their nominal value.

5.6.3 The Process Being Modeled and Data

The initially envisaged use of the computer model was to replace (or massively reduce) the need for the field-testing of actual vehicles on a test track that included several stressors (potholes). The result of a vehicle test is a

---

[3] See http://www.mscsoftware.com/Products/CAE-Tools/Adams.aspx. Accessed September 1, 2011.

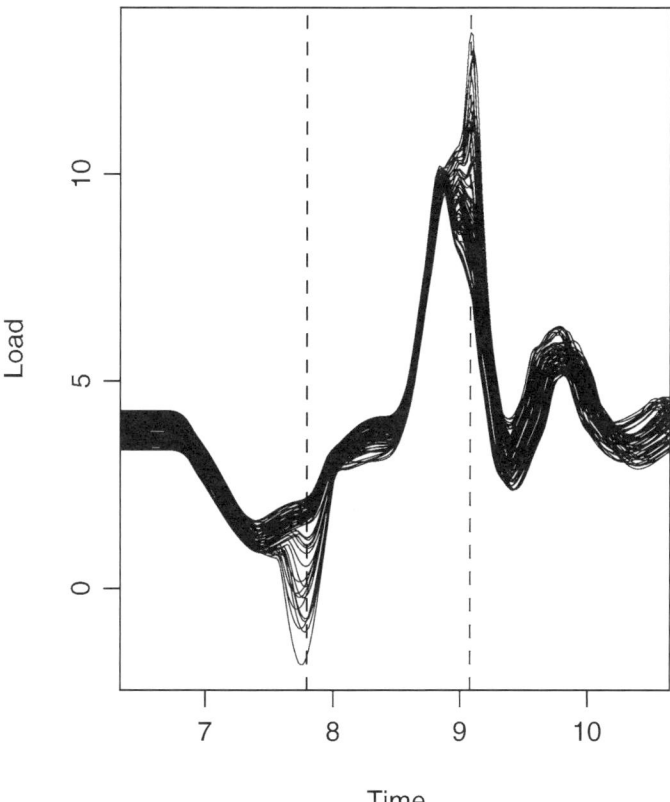

FIGURE 5.3 Computer model predictions of force on the suspension system at 65 inputs. SOURCE: Bayarri et al. (2007a).

time trace of the load on the suspension system as the vehicle drives down the test track. Figure 5.3 presents the computer model prediction of these time traces for 65 different combinations of values for the nine input parameters described above. These 65 input value sets were chosen, using a Latin hypercube design, so as to "cover" the design space of possible input values in a representative fashion. (For simplicity, only part of the time traces are given in Figure 5.3—as the vehicle runs over a single pothole and a span of about 3 m—and only analyses for this region are discussed here.)

This was a context in which the predictive approach to model validation could be entertained; the validation process would center on field data obtained from actual physical runs of a vehicle over the test track. A test vehicle was outfitted with sensors at various locations on the suspension system and was physically driven seven times over the test track. This resulted in seven "real" independent time series of road loads, measured subject to random error but not to bias.

5.6.4 Modeling the Uncertainties

To understand the uncertainties in predictions of the computer model, it is first necessary to model the uncertainties in model inputs, the real-process data, and the model itself. For the nine model input parameters, these uncertainties were given in the form of prior probability distributions, obtained by consultation with the engineers involved with the project. Many of these were simply the known distributions of suspension system parts arising from manufacturing variability. The measurement errors in the data were modeled using a wavelet decomposition process.

There were two sources of uncertainty concerning the computer model itself. The model discrepancy issue discussed in Section 5.3, "Model Calibration and Inverse Problems," was handled by allowing a functional deviation of the computer model from reality and a Gaussian process prior to this discrepancy (following Kennedy and O'Hagan, 2001). The second source of uncertainty in the model was in the use of an emulator (an approximation to the computer model) because of the expense of running the computer model. Since a Gaussian process was used to construct the emulator, the uncertainty in its approximation can be readily incorporated into the overall assessment of uncertainty.

5.6.5 Analysis and Results

The collection of probability distributions representing the uncertain model inputs, uncertain real-process data, and uncertain model was processed through Bayesian analysis involving MCMC computation (Bayarri et al., 2007b). The results of the analysis are expressed as posterior distributions of QOIs, summarized by a posterior expected values (the "prediction" of the quantity) and confidence intervals to indicate the uncertainty in the prediction.

Figure 5.4 presents the estimated model discrepancy—that is, the estimated difference between the computer-model prediction and the real process. The dashed line is the mean discrepancy, and the solid lines are 90 percent

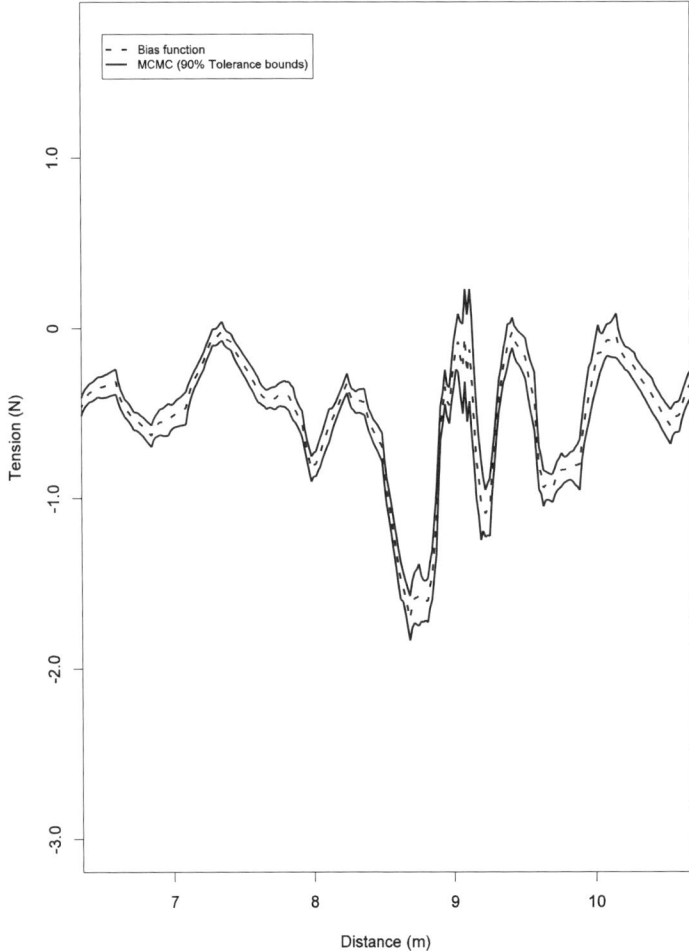

FIGURE 5.4 Estimated discrepancy of the computer model from reality. SOURCE: Bayarri et al. (2007b).

uncertainty bands for this discrepancy. As can be seen, the computer model provided a reasonable approximation to the real process; a constant value of zero for the discrepancy would indicate a perfect validation.

The shown deviation from zero was not deemed by the engineers to be excessive, except for the spike (which occurred at the pothole). An effort was made to incorporate the discrepancy assessment into improved predictions of reality from the computer model. Note that this incorporation is not a matter of simply "adding" the discrepancy to the model prediction because the uncertainties in the discrepancy and damping parameters are highly dependent, so that their joint posterior distribution must be utilized to determine the adjustment.

Various prediction scenarios can be considered. The most challenging of these was the prediction for a new vehicle type having different nominal values of the input parameters. The most valuable engineering use of computer models is, indeed, to extrapolate to such a system having new input values, since then the expense of obtaining real-process data can be avoided. Such extrapolation requires strong assumptions concerning the discrepancy. The simplest assumption is that the new system has the same discrepancy function (or distribution) as the old system, and that was assumed here. Because of physical understanding of the system, the discrepancy was viewed as being multiplicative rather than additive.

To predict the road-load time trace for the new vehicle type, the computer model (in which the new vehicle type was close enough in design to allow the use of the same finite-element representation as for the previous vehicle type) was run at 65 values of the uncertain inputs. These were now centered around nominal input values assessed for the new vehicle. A Bayesian analysis was subsequently performed, utilizing these 65 computer-model runs, the prior distributions for the seven suspension characteristics, and the joint posterior distribution of the discrepancy and damping parameters obtained from the original vehicle.

Figure 5.5 summarizes the results of this analysis. Clearly there is a significant difference between the predictions of the computer model alone and the predictions that incorporated the discrepancy of the model that was determined by using previous experimental data. Eventually, the new vehicle was driven over the test track, with the actual road-load traces being measured—the heavy black band in Figure 5.5. While not perfect, the Bayesian predictions are considerably more accurate than predictions from the computer model alone.

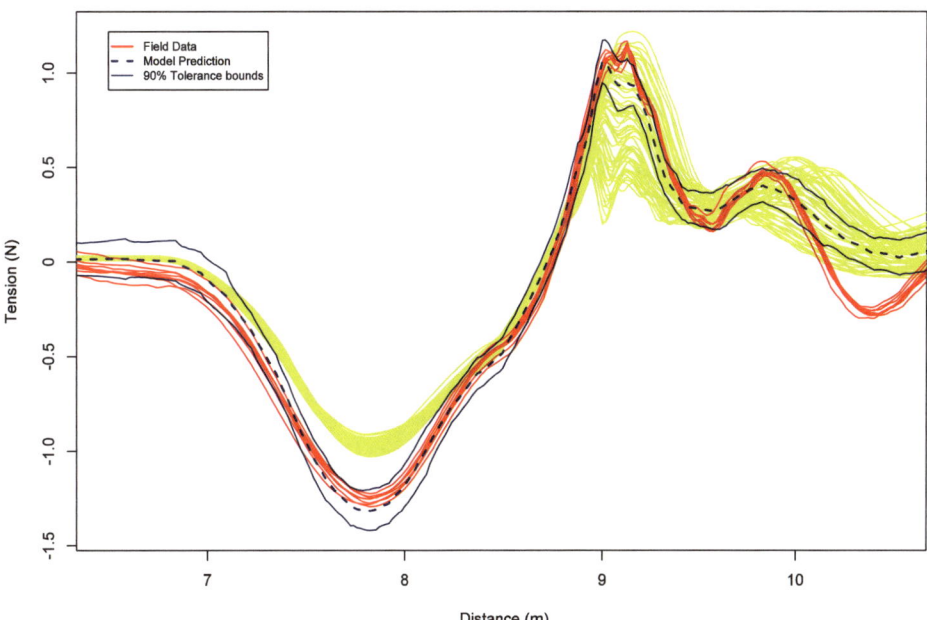

FIGURE 5.5 Results of analysis. The green curves show the 65 computer-model runs. The Bayesian posterior mean of the road-load trace is the dashed line, the 90 percent uncertainty bands are the solid blue lines, and the heavy red line is the field data. SOURCE: Bayarri et al. (2007b).

This example shows a successful use of discrepancy-adjusted prediction; such adjustment will not always be acceptable, however, because extrapolation beyond the range of relevant data is always a challenge, depending as it does on assumptions about continuity in the underlying model validity and about conditions under which extrapolating estimates of parameter values can be assumed to be smooth. However, since the computer model itself is used to do the computational "heavy lifting" in the extrapolation, with the discrepancy adjustment playing a lesser role, concerns about extrapolation can be somewhat mitigated.

More important, the Bayesian approach implicitly incorporates all uncertainties in the analysis and folds these into assessment of the overall uncertainty in the posterior distributions of all QOI. The main limitation of the approach is the need for MCMC computation, which could sometimes require either extensive computational effort or the development of a good emulator for the more computationally challenging components of the computer model.

5.7 INFERENCE FROM MULTIPLE COMPUTER MODELS

In applications such as climate change, uncertainties in 20- or 100-year forecasts are likely dominated by structural uncertainty—uncertainty due to the discrepancy between model and reality. Since there are few or no physical observations from which to estimate model discrepancy directly, predictions from a number of different climate models are often used to help quantify prediction uncertainty.

Some forecasting applications have made successful use of combining predictions from different models (e.g., Gneiting and Raftery, 2005). These approaches incorporate multiple model-based predictions within a statistical modeling framework, often producing predictions more accurate those of than any single model, with more-reliable estimated uncertainties. Although climate and weather are the main examples discussed in this section, these basic approaches have been applied in a wider variety of applications.

The Intergovernmental Panel on Climate Change (IPCC) uses predictions of the future climate state from multiple computational global climate models, developed under various research efforts from around the world, to make assessments about climate change under different future emission scenarios (Meehl et al., 2007). The resulting multimodel ensemble (MME) of climate model runs, carried out largely for the IPCC assessment, has been used by a variety of researchers to produce predictions of future climate along with estimates of prediction uncertainty. The most common approach for using such an MME to make predictions about future climate is the use of a hierarchical modeling framework, effectively treating the output from different models as noisy versions of the actual climate system (Tebaldi et al., 2005; Buser et al., 2009; Smith et al., 2010). Researchers readily admit that such hierarchical modeling approaches are far from ideal—the ensemble of models is a sample of convenience, and dependencies between different computational models are typically not accounted for in the statistical modeling. Interestingly, the estimated prediction uncertainty typically decreases as the ensemble size increases. It is not at all clear that this should be the case. One can argue that even if infinitely many models could be sampled, future climate would still not be perfectly understood because of our limited knowledge of climate physics. However, the hierarchical modeling approach is a first step for developing more realistic prediction uncertainties using MMEs. Research is needed to establish the connection between model-to-model differences and model-to-reality differences.

In forecasting applications, where repeated relevant physical observations are abundant, the combination of multiple computational models has led to improved predictions and improved prediction uncertainties. The probcast (web page: http://probcast.washington.edu/) methodology of Gneiting and Raftery (2005) is a notable example. This approach, like many others, uses Bayesian model averaging (Hoeting et al., 1999), which models the physical observations as coming from one of the models in the ensemble—but which model is chosen is uncertain. The resulting analysis produces a posterior distribution for the forecast that is a weighted average of the individual model predictions (with uncertainty). The resulting predictions from the Bayesian model averaging approach are generally more accurate than any single prediction, and the resulting prediction uncertainty better describes the variation of the prediction about the observed value than does the raw ensemble.

So far, the success of such model-averaging approaches in forecasting has not been translated to more extrapolative, data-poor settings such as climate change where predictions and their uncertainty cannot be calibrated with an abundant supply of relevant physical observations. The hierarchical model-based approaches for assessing prediction uncertainty using multimodel ensembles can deal with the relative paucity of physical observations and can capture key sources of uncertainty that may be missed using more traditional parametric variations within a single computational model, but are justified only under assumptions that are often not met in practice. Additional research will likely improve the state of the art in combining predictions from multimodel ensembles. Such research includes improved methods for constructing ensembles of models, analysis of interdependence among models, assessment of confidence in particular models and their predictive power, and use of information-theoretic and statistical means for developing robust and reliable methods for model comparison, selection, and averaging/pooling.

5.8 EXPLOITING MULTIPLE SOURCES OF PHYSICAL OBSERVATIONS

In many applications, multiple sources of physical observations may be available for the validation/prediction assessment. In engineering applications, the data sources might conform to a validation hierarchy (see Figure 5.7 in Section 5.9.5), whereas in other applications these different data sources might include different sensing modalities (e.g., infrared, visible, seismic) or different data sources (e.g., pressure measurements or well cores). It may also be appropriate to use output from high-quality simulations as surrogates for physical observations (e.g., direct numerical simulation of turbulent flow using resolved Navier-Stokes equations may inform about predictions using coarser, Reynolds-averaged, Navier-Stokes simulations). There is the opportunity to make use of these various sources of physical observations to address key issues such as model calibration, model discrepancy, prediction uncertainty, and assessing the quality of the prediction. There is also opportunity to use what is learned from such analyses to inform how to select additional observations or to design additional experiments.

For a given collection of physical observations, there is the question of how best to use these sources for validation and prediction. For example, should low-level experiments in a validation hierarchy be used for calibration, saving the more integrated experiments for assessing the model? Or should both calibration and assessment be done together? Different strategies will require different approaches, which may affect the quality of the predictions.

Multiple sources of physical observations provide an opportunity to assess a prediction and the accompanying prediction uncertainty. One way to exploit this opportunity is to identify collections of experiments, or observation sources, that can be used to assess the quality of a "surrogate" prediction that has important commonalities with the QOI prediction. The characteristics that define an appropriate surrogate, if they exist, will depend on features of the domain space. Does a candidate surrogate prediction depend on the physical process in a way similar to the QOI prediction? Does the surrogate have similar sensitivities to model inputs? Does the model discrepancy function (if there is one) adequately capture uncertainty for these predictions? Should the same model discrepancy function transfer to the QOI? Exactly how best to use multiple sources of physical data to improve the quality and accuracy of predictions is an active VVUQ research area.

In cases where the validation effort will call for additional experiments, the methodologies of validation and prediction can be used to help assess the value of additional experiments and might also suggest new types of experiments to address weaknesses in the assessment. Ideas from the design of experiments from statistics (Wu and Hamada, 2009) are relevant here, but the design of validation experiments involves additional complications that make this an open research topic. The computational demands of the computational model are a complicating factor, as is the issue of dealing with model discrepancies. Also, some of the key requirements for additional experiments—such as improving the reliability of the assessment or improving communication to stakeholders or decision makers—are not easily quantified. The experimental planning enterprise is considered from a broader perspective in Chapter 6.

5.9 PECOS CASE STUDY

5.9.1 Overview

The Center for Predictive Engineering and Computational Sciences, called the PECOS Center, at the University of Texas at Austin is part of the Predictive Science Academic Alliance Program (PSAAP) of the Department of Energy's National Nuclear Security Administration. The PECOS Center is engaged in developing VVUQ processes to gain an understanding of the reentry of a space capsule (e.g., NASA's proposed Orion vehicle) into Earth's atmosphere. Of primary interest is the performance of the thermal protection system (TPS), which protects the vehicle from the extreme thermal environment arising from travel through the atmosphere at speeds of Mach 20 or higher, depending on the trajectory. Vehicles that use ablative heat shields (e.g., Orion and Apollo) are being simulated to predict the rate at which the ablator is being consumed.

TPS consumption is a critical issue in the design and operation of a reentry vehicle—if the entire heat shield is consumed, the vehicle will burn up. TPS consumption is governed by a range of physical phenomena, including high speed and turbulent fluid flow, high-temperature aero-thermo-chemistry, radiative heating, and the response of complex materials (the ablator). Thus, a numerical simulation of reentry vehicles requires models of these phenomena.

The reentry vehicle simulations share a number of complicating characteristics with many other high-consequence computational science applications. These complicating characteristics include the following:

- The QOIs are not accessible for direct measurement under the conditions in which the predictions are to be made;
- The predictions involve multiple interacting physical models;
- Experimental data available for calibrating and validating models are difficult to obtain, include significant uncertainty, are sparse, and often describe physical conditions not directly related to the predictions; and
- The best-available models for some of the physical phenomena are known to include sizable errors.

These characteristics greatly complicate the assessment of prediction reliability and the application of VVUQ techniques.

5.9.2 Verification

As described above, there are two components of the verification of computer simulation: (1) ensuring that the computer code used in the simulation correctly implements the intended numerical discretization of the model (code verification) and (2) ensuring that the errors introduced by the numerical discretization are sufficiently small (solution verification).

5.9.3 Code Verification

There are many aspects of ensuring the correct implementation of a mathematical model in a computer code. Many of these are just good software engineering practices, such as exhaustive model development and user documentation, modern software design, configuration control, and continuous unit and regression testing. Commonly understood to be important but less commonly practiced, these processes are an integral part of the PECOS software environment.

To ensure that an implementation is actually producing correct solutions, one wants to compare results to known, preferably analytic, solutions. Unfortunately, analytic solutions are not generally available, which is the reason for the use of the *method of manufactured solutions* (MMS), in which source terms are added to the equations to make a prespecified "solution" exact (Steinberg and Roache, 1985; Roache, 1998; Knupp and Salari, 2003; Long et al., 2010; and Oberkampf and Roy, 2010). Although MMS is a widely recognized approach, it is not commonly used. One reason is that it is much more difficult to implement for complex problems than it appears. First, even for systems of moderate complexity (e.g., three-dimensional compressible Navier-Stokes) there can be many hundreds of source terms, and it is clearly necessary that the evaluation of these terms be done with high reliability. Thus, constructing analytic solutions is itself a software engineering and reliability challenge. Second, the introduction of the source terms into the code being tested must be done with minimal (preferably no) changes to the code, so that the tests are relevant to the code as it will be used. Unfortunately, this introduction of the source terms may not be possible in codes that have not been designed for it. Finally, it is necessary that manufactured solutions have characteristics similar to those of the problems that the codes will be used to solve. This is important so that bugs are not masked by the fact that the terms in which they occur may be insignificant in a manufactured solution that is too simple.

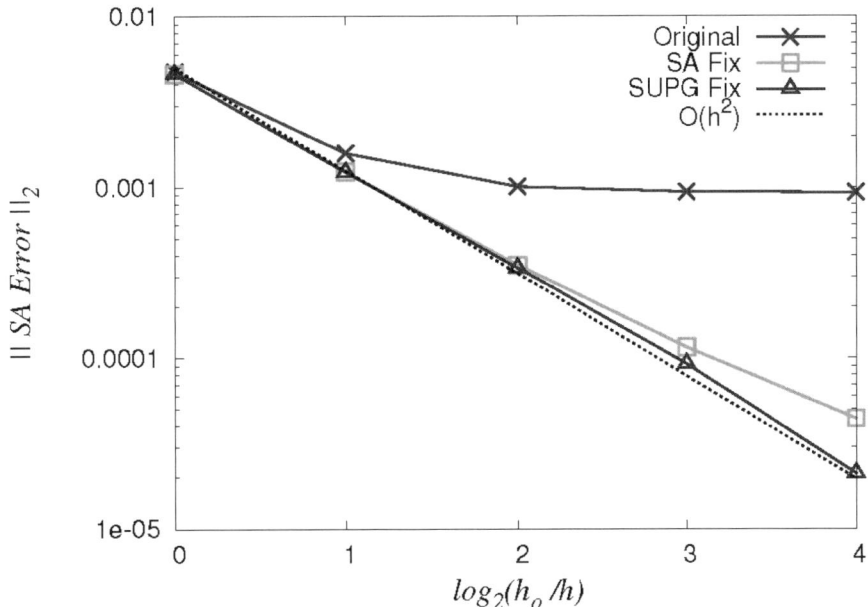

FIGURE 5.6 Dependence of the L_2 error in the Spalart-Allmaras (SA) turbulence model manufactured solution on the grid size, under uniform refinement. Shown are the original test, the test after the correction of a bug in the SA equations, and that after subsequent correction of a bug in streamline-upwind/Petrov Galerkin regularization. The theoretical convergence is second order.

At the PECOS Center, to make MMS useful for the verification of reentry vehicle codes, a highly reliable software library for implementing manufactured solutions (the Manufactured Analytic Solution Abstraction, or MASA) and a library of manufactured solutions, using symbolic manipulation software (e.g., Maple), have been developed. These manufactured solutions have been imported into MASA. MASA and associated solutions have been publicly released.[4] Further, part of the PECOS Center software development process involves developing and documenting a verification plan (usually involving MMS) before development begins, so that codes are designed to enable MMS. These efforts have paid off by exposing a number of subtle but important bugs in PECOS software. An example is shown in Figure 5.6, in which the convergence with grid refinement of the Spalart-Allmaras (SA) turbulence model[5] equations to a manufactured solution is shown. In the initial test, the solution error did not converge to zero with uniform grid refinement, which led to the discovery of a bug in the implementation of the SA equations. When this bug was fixed, the error did reduce with refinement, but not at the theoretically expected rate of h^2. The slower rate was caused by a long-standing bug in the implementation of streamline-upwind/Petrov-Galerkin (SUPG) stabilization in the LibMesh finite-element infrastructure in which the model was implemented.

5.9.4 Solution Verification

The question in solution verification is whether a numerical solution to a set of model equations is "close enough" to the exact solution. A "close enough" standard is necessary because, although discretization errors can generally be made arbitrarily small through refinement of the discretization, it is neither practical nor necessary to drive these errors to the level of round-off error. Generally, the models are used to predict certain output QOIs, and one wants to ensure that these quantities are within some tolerance of those from the exact solution of the models.

---

[4] See https://red.ices.utexas.edu/projects/software/wiki/MASA. Accessed March 19, 2012.
[5] For a definition and details of this model, see http://turbmodels.larc.nasa.gov/spalart.html. Accessed March 24, 2012.

An acceptable numerical error tolerance depends on the circumstances. At the PECOS Center, since the numerical discretization errors are under the control of the analyst, the view is taken that they should be made sufficiently small to be negligible compared to other sources of uncertainty. This avoids the need to model the uncertainty arising from such errors. It is important to identify the QOIs for which predictions are being made because the numerical discretization requirements for predicting some quantities (e.g., high-order derivatives) are much more stringent than for other quantities.

Solution verification, then, requires that the discretization error in the QOIs be estimated. The common practice of comparing solutions on two grids to check how much they differ is not sufficient. In simple situations it is possible to refine the discretization uniformly (e.g., half the grid spacing everywhere) and then to apply Richardson extrapolation to develop an error estimate. A more general technique, and the one used at the PECOS Center, is adjoint-based a posteriori error estimation (Bangerth and Rannacher, 2003). Once one has an estimate of the errors in the QOIs, it may be necessary to refine the discretization to reduce this error. Adjoint-based error estimators also provide an indicator of where (in space and/or time) the discretization errors are contributing most to errors in the QOIs. Goal-oriented adaptivity (Bangerth and Rannacher, 2003; Oden and Prudhomme, 1998; Prudhomme and Oden, 1999; Strouboules et al., 2000) uses this adjoint information to drive adaptive refinement of the discretization.

At the PECOS Center, the simulation codes used to make predictions of the ablator consumption rate (the QOI) have been developed to perform adjoint-based error estimation and goal-oriented refinement. For example, a hypersonic flow code (FIN-S) supporting goal-oriented refinement was built on the LibMesh infrastructure (Kirk et al., 2006). Adaptivity is used to reduced the estimated error in the QOIs to below specified tolerances, thereby accomplishing solution verification.

5.9.5 Validation

Data and associated models of data uncertainty are critical to predictive simulation. They are needed for the calibration of physical models and inadequacy models and for the validation of these models. At the PECOS Center, the calibration, validation, and prediction processes are closely related, interdependent, and at the heart of uncertainty quantification in computational modeling.

A number of complications arise from the need to pursue validation in the context of a QOI. First, note that in most situations the QOI in the prediction scenario is not accessible for observation, since otherwise, a prediction would generally not be needed. This inability to observe the QOI can arise for many reasons, such as legal or ethical restrictions, lack of instrumentation, limitations of laboratory facilities to reproduce the prediction scenario, cost, or that the prediction is about the future. At the PECOS Center, the QOI is the consumption rate of an ablative heat shield at peak heating for a particular trajectory of a reentry vehicle. It is experimentally unobservable because the conditions are not accessible in the laboratory and because flight tests are expensive, making it impractical to test every trajectory of interest.

Validation tests are of course posed by comparing to observations the outputs of the model for some observable quantity. The central challenge is to determine what the mismatch between observations and the model, and the relevant prediction uncertainties, imply about predictions of unobserved QOIs. Because the QOIs cannot be observed, the only access that one has to them is through the model, and so this assessment can be done only in the context of the model.

Another complication arises when the system being modeled is complicated with many parts or encompasses many interacting physical phenomena. In this case, the validation process is commonly hierarchical, with validation tests of models for subcomponents or individual physical phenomena based on relatively simple (inexpensive) experiments. As an example, in the reentry vehicle problem being pursued at the PECOS Center, the individual physical phenomena include aero-chemistry, turbulence, thermal radiation, surface chemistry, and ablator material response.

Combinations of subcomponents or physical phenomena are then tested against more complicated, less-abundant multiphysics experiments. Finally, in the best circumstances, one has some experimental observations available for the complete system, allowing a validation test for the complete model. The hierarchical validation process can be envisioned as a validation pyramid shown in Figure 5.7.

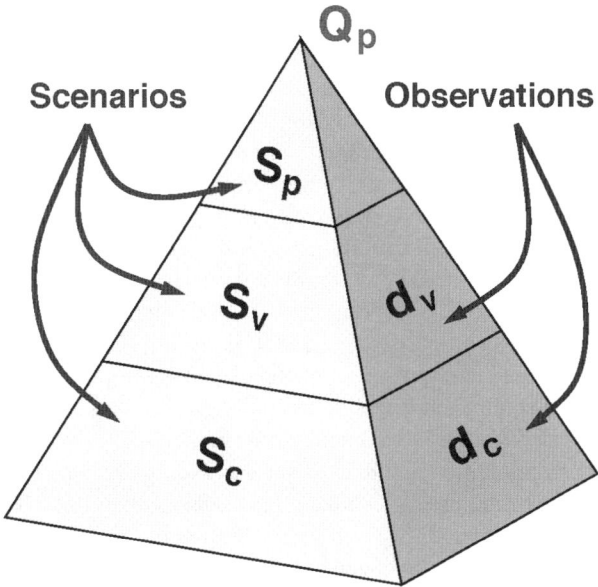

FIGURE 5.7 The prediction pyramid depicting the increasing complexity of the physical scenarios (S_c, S_v, and S_p) accompanied by the decreasing availability of data (d_c and d_v) for calibration and validation of complex multiphysics models, with the prediction quantity of interest (Q_p) residing at the highest level of the pyramid.

The hierarchical nature of multiphysics validation poses further challenges. First, the QOIs are generally accessible only through the model of the full system, so that single-physics models do not have access to the QOI, making QOI-aware validation difficult. Generally, surrogate QOIs are devised for single-physics models—a surrogate QOI being as closely related to the full system QOI as possible. For example, in the validation of boundary-layer turbulence models for the reentry vehicle simulations pursued at the PECOS Center, the turbulent wall heat flux is identified as a surrogate QOI, since it is directly related to, and is a driver for, the ablation rate. Multiphysics validation tests performed at higher levels of the pyramid are important because they generally test the models for the coupling between the single-physics models. But the fact that data are generally scarce at these higher levels means that these coupling models are commonly not as rigorously tested as the simpler models are, affecting the overall quality of the final prediction.

5.10 RARE, HIGH-CONSEQUENCE EVENTS

Large-scale computational models play a role in the assessment and mitigation of rare, high-consequence events. By definition, such events occur very infrequently, which means that there is little measured data from them. Thus, the issues that complicate extrapolative predictions are almost always present in predictions involving rare events. Still, computational models play a key role in safety assessments for nuclear reactors by the Nuclear Regulatory Commission (Mosleh et al., 1998) and in assessing safety risks in subsurface contaminant transport at Department of Energy facilities (Neuman and Wierenga, 2003). Computational models also play a role in characterizing the causes and consequences of potential natural disasters such as earthquakes, tsumanis, severe storms, avalanches, fires, or even meteor impacts. The behavior of engineered systems (e.g., bridges, buildings) under extreme conditions, or simply as a result of aging and normal wear and tear, can also fall under this heading of rare, high-consequence events.

In many cases, such as probabilistic risk assessment (Kumamoto and Henley, 1996) applied to nuclear reactor safety, computational models are used to evaluate the consequences of identified scenarios, helping to quantify the risk—the product of the chance of an event and its consequences. This is also true of assessments of the risks from large meteor impacts, for which computer models simulate the consequences of impacts under different conditions (Furnish et al., 1995). Although it is difficult to assess confidence in such extrapolative predictions, their results can be integrated into a larger risk analysis to prioritize threats. In such analyses, it may be a more efficient use of resources to further scrutinize the model results only for the threats with highest priority.

Computational models can also be used to seek out combinations of initial conditions, forcings, and even parameter settings that give rise to extreme, or high-consequence, events. Assessing the chances of such events comes after their discovery. Many of the methods described in Chapters 3 and 4 are relevant to this task, but now with a focus on finding aberrant behavior rather than inferring settings that match measurements. This may involve exploring how a physical system can be "stretched" to produce (as yet) unseen, extreme behavior, perhaps induced by interactions among different processes. This is the opposite of designing, or engineering, a system to ensure that interactions among the various processes are minimized. Calculating such extreme behavior may tax a model to the point that its ability to reproduce reality is questionable. Methods for assessing and improving confidence in such model predictions are challenging and largely open problems, as they are for extrapolative predictions.

Once a high-consequence event is identified, computational models can be viable tools for assessing its probability. Such events are rare, and so standard approaches such as Monte Carlo simulation are infeasible because large numbers of model runs would be required to estimate these small probabilities. There are rich lines of current research in this area. Oakley and O'Hagan (2004) use a combination of emulation and importance sampling for assessing small probabilities in infrastructure management. Picard (2005) biases a particle-based code to produce more extreme events, statistically adjusting for this bias in producing estimates. In addition to response-surface approaches, one might also use a combination of high- and low-fidelity models to seek out and estimate rare-event probabilities. Another possible multifidelity strategy would be to use a low-fidelity model to seed promising boundary conditions to a high-fidelity, localized model (Sain et al., 2011). Embedding computational models in standard statistical approaches is another promising direction. For example, Cooley (2009) combines computer model output and extreme value theory from statistics to estimate the frequency of extreme rainfall events. Bayarri et al. (2009b) utilize a computer model to identify the catastrophic region in input space for extreme pyroclastic volcanic flows and statistical modeling of the input distributions to compute the probability of the extreme events.

A better understanding of complex dynamical systems could help in the search for precursors to extreme events or important changes in system dynamics (Scheffer et al., 2009). Computational models will likely have a role in such searches—even when the models are known to have shortcomings in their representation of such complex systems. Currently, computational models are being used to help inform monitoring efforts, helping to provide early warnings of events ranging from groundwater contamination to a terrorist attack.

Finally, bounding and "worst-case" approaches, if not too conservative, can provide actionable information about rare, high-consequence events. Recent work by Lucas et al. (2009) uses concentration-of-measure inequalities to bound the probability of extreme outcomes, without having to specify fully the distribution of the input uncertainties. Also, more traditional decision-theoretic approaches (e.g., minimax decision rules [Berger, 1985]; worst-case priors [Evans and Stark, 2002]) may be useful for dealing with rare, high-consequence events. One could imagine embedding these ideas into a computational model, using a worst-case value for a reaction coefficient, a permeability field, a boundary condition, or even how a physical process is represented in the computational model.

5.11 CONCLUSION

This chapter discusses numerous tasks that contribute to validation and prediction from the perspective of mathematical foundations, pointing out areas of potential fruitful research. As noted, details of these tasks depend substantially on the features of the application—the maturity, quality, and speed of the computational model; the available physical observations; and their relation to the QOI. The concept of embedding the computational model within a mathematical/statistical framework that can account for and model relevant uncertainties, including those caused by initial and boundary conditions, input parameters, and model discrepancy is also described.

Some applications involve making predictions and uncertainty estimates in settings for which physical observations are plentiful. In even mildly extrapolative settings, obtaining these estimates and assessing their reliability remains an open problem. The NRC (2007) report on the use of models in environmental regulatory decision making states, "When model results are to be extrapolated outside of conditions for which they have been evaluated, it is important that they have the strongest possible theoretical basis, explicitly representing the processes that will most affect outcomes in the new conditions to be modeled, and embodying the best possible parameter estimates" (p. 129). The findings and recommendation below relate to making extrapolative predictions.

Finding: Mathematical considerations alone cannot address the appropriateness of a model prediction in a new, untested setting. Quantifying uncertainties and assessing their reliability for a prediction require both statistical and subject-matter reasoning.

Finding: The idea of a *domain of applicability* is helpful for communicating the conditions for which predictions (with uncertainty) can be trusted. However, the mathematical foundations have not been established for defining such a domain or its boundaries.

Finding: Research and development on methods for assessing uncertainties of model-based predictions in new, untested conditions (i.e., "extrapolations") will likely require expertise from mathematics, statistics, computational modeling, and the science and engineering areas relevant to a given application. Specific needs in assessing uncertainties in prediction include:

- Approaches for specifying and estimating model discrepancy terms that leverage physical understanding, features of the application, and known strengths and deficiencies of the computational model for the application;
- Computational models developed with VVUQ in mind, which might include the need for availability of derivative information; a faster, lower-fidelity representation of the model (perhaps with specified discrepancy); or embedding physically motivated discrepancy terms within the model that can produce more reliable prediction uncertainties for the QOI and that can be calibrated with available physical observations;
- A framework for efficiently exploiting a hierarchy of available experiments—allocating experiments for calibration, assessing prediction accuracy, assessing the reliability of predictions, and suggesting new experiments within the hierarchy that would improve the quality of estimated prediction uncertainties;
- Guidelines for reporting predictions and accompanying prediction uncertainties, including disclosure of which sources of uncertainty are accounted for, which are not, what assumptions these estimates rely on, and the reliability or quality of these assumptions; and
- Compelling examples of VVUQ done well in problems with different degrees of complexity.

A similar conclusion was reached by the National Science Foundation (NSF) Division of Mathematics and Physical Sciences (MPS), which in its May 2010 advisory committee report recommended as follows:

> MPS should encourage interdisciplinary interaction between domain scientists and mathematicians on the topic of uncertainty quantification, verification and validation, risk assessment, and decision making. (NSF, 2010)

The above ideas are particularly relevant to the modeling of complex systems where even a slight deviation from physically tested conditions may change features of the system in many ways, some of which are incorporated in the model and some of which are not.

The field of VVUQ is still developing, making it too soon to offer any specific recommendations regarding particular methods and approaches. However, a number of principles and accompanying best practices are listed below regarding validation and prediction from the perspective of mathematical foundations.

- Principle: A validation assessment is well-defined only in terms of specified QOIs and the accuracy needed for the intended use of the model.
    —Best practice: Early in the validation process, specify the QOIs that will be addressed and the required accuracy.
    —Best practice: Tailor the level of effort in assessment and estimation of prediction uncertainties to the needs of the application.

- Principle: A validation assessment provides direct information about model accuracy only in the domain of applicability that is "covered" by the physical observations employed in the assessment.
  - Best practice: When quantifying or bounding model error for a QOI in the problem at hand, systematically assess the relevance of supporting data and validation assessments (which were based on data from different problems, often with different QOIs). Subject-matter expertise should inform this assessment of relevance (as discussed above and in Chapter 7).
  - Best practice: If possible, use a broad range of physical observation sources so that the accuracy of a model can be checked under different conditions and at multiple levels of integration.
  - Best practice: Use "holdout tests" to test validation and prediction methodologies. In such a test some validation data are withheld from the validation process, the prediction machinery is employed to "predict" the withheld QOIs, with quantified uncertainties, and finally the predictions are compared to the withheld data.
  - Best practice: If the desired QOI was not observed for the physical systems used in the validation process, compare sensitivities of the available physical observations with those of the QOI.
  - Best practice: Consider multiple metrics for comparing model outputs against physical observations.
- Principle: The efficiency and effectiveness of validation and prediction assessments are often improved by exploiting the hierarchical composition of computational and mathematical models, with assessments beginning on the lowest-level building blocks and proceeding to successively more complex levels.
  - Best practice: Identify hierarchies in computational and mathematical models, seek measured data that facilitate hierarchical validation assessments, and exploit the hierarchical composition to the extent possible.
  - Best practice: If possible, use physical observations, especially at more basic levels of the hierarchy, to constrain uncertainties in model inputs and parameters.
- Principle: Validation and prediction often involve specifying or calibrating model parameters.
  - Best practice: Be explicit about what data/information sources are used to fix or constrain model parameters.
  - Best practice: If possible, use a broad range of observations over carefully chosen conditions to produce more reliable parameter estimates and uncertainties, with less "trade-off" between different model parameters.
- Principle: The uncertainty in the prediction of a physical QOI must be aggregated from uncertainties and errors introduced by many sources, including discrepancies in the mathematical model, numerical and code errors in the computational model, and uncertainties in model inputs and parameters.
  - Best practice: Document assumptions that go into the assessment of uncertainty in the predicted QOI, and also document any omitted factors. Record the justification for each assumption and omission.
  - Best practice: Assess the sensitivity of the predicted QOI and its associated uncertainties to each source of uncertainty as well as to key assumptions and omissions.
  - Best practice: Document key judgments—including those regarding the relevance of validation studies to the problem at hand—and assess the sensitivity of the predicted QOI and its associated uncertainties to reasonable variations in these judgments.
  - Best practice: The methodology used to estimate uncertainty in the prediction of a physical QOI should also be equipped to identify paths for reducing uncertainty.
- Principle: Validation assessments must take into account the uncertainties and errors in physical observations (measured data).
  - Best practice: Identify all important sources of uncertainty/error in validation data—including instrument calibration, uncontrolled variation in initial conditions, variability in measurement setup, and so on—and quantify the impact of each.
  - Best practice: If possible, use replications to help estimate variability and measurement uncertainty.
  - Remark: Assessing measurement uncertainties can be difficult when the "measured" quantity is actually the product of an auxiliary inverse problem—that is, when it is not measured directly but is inferred from other measured quantities.

Finally, it is worth pointing out that there is a fairly extensive literature in statistics focused on model assessment that may be helpful if adapted to the model validation process. Basic principles such as model diagnostics (Gelman et al., 1996; Cook and Weisberg, 1999), visualization and graphical methods (Cleveland, 1984; Anselin, 1999), hypothesis testing and model selection (Raftery, 1996; Bayarri and Berger, 2000; Robins et al., 2000; Lehmann and Romano, 2005), cross-validation and the use of holdout tests (Hastie et al., 2009) could play central roles in validation and prediction, as they do for statistical model checking.

5.12 REFERENCES

Akçelik, V., G. Biros, A. Draganescu, O. Ghattas, J. Hill, and B. Van Bloeman Waanders. 2005. Dynamic Data-Driven Inversion for Terascale Simulations: Real-Time Identification of Airborne Contaminants, in *Proceedings of SC2005*.

AIAA (American Institute for Aeronautics and Astronautics). 1998. *Guide for the Verification and Validation of Computational Fluid Dynamics Simulations*. Reston, Va.: AIAA.

Anselin, L., 1999. Interactive Techniques and Exploratory Spatial Data Analysis. *Geographical Information Systems: Principles, Techniques, Management and Applications* 1:251-264.

Badri Narayanan, V.A., and N. Zabaras. 2004. Stochastic Inverse Heat Conduction Using a Spectral Approach. *International Journal for Numerical Methods Engineering* 60:1569-1593.

Bangerth, W., and R. Rannacher. 2003. *Adaptive Finite Element Methods for Differential Equations*. Basil, Switzerland: Birkhauser Verlag.

Bayarri, M.J., and J.O. Berger. 2000. P Values for Composite Null Models. *Journal of the American Statistical Association* 95(452):1269-1276.

Bayarri, M.J., J.O. Berger, M.C. Kennedy, A. Kottas, R. Paulo, J. Sacks, J.A. Cafeo, C.H. Lin, and J. Tu. 2005. *Bayesian Validation of a Computer Model for Vehicle Crashworthiness*. Technical Report 163. Research Triangle Park, N.C: National Institute of Statistical Sciences.

Bayarri, M.J., J. Berger, R. Paulo, J. Sacks, J. Cafeo, J. Cavendish, C. Lin, and J. Tu. 2007a. A Framework for Validation of Computer Models. *Technometrics* 49:138-154.

Bayarri, M.J., J. Berger, G. Garcia-Donato, F. Liu, J. Palomo, R. Paulo, J. Sacks, D. Walsh, J. Cafeo, and R. Parthasarathy. 2007b. Computer Model Validation with Functional Output. *Annals of Statistics* 35:1874-1906.

Bayarri, M.J., J.O. Berger, M.C. Kennedy, A. Kottas, R. Paulo, J. Sacks, J.A. Cafeo, C.H. Lin, and J. Tu. 2009a. Predicting Vehicle Crashworthiness: Validation of Computer Models for Functional and Hierarchical Data. *Journal of the American Statistical Association* 104:929-943.

Bayarri, M.J., J.O. Berger, E.S. Calder, K. Dalbey, S. Lunagomez, A.K. Patra, E.B. Pitman, E.T. Spiller, and R.L. Wolpert. 2009b. Using Statistical and Computer Models to Quantify Volcanic Hazards. *Technometrics* 5:402-413.

Berger, J.O. 1985. *Statistical Decision Theory and Bayesian Analysis*. New York: Springer.

Berger, J.L., and R.L. Wolpert. 1988. The Liklihood Principle. Lecture notes available at http://books.google.com/books?hl=en&lr=&id=7fz8JGLmWbgC&oi=fnd&pg=PA1&dq=berger+and+wolpert+the+likelihood+principle&ots=iTkq2Ekz_Z&sig=qKnLby2avTKEP_unAWSJ_BUI#v=onepage&q=berger%20and%20wolpert%20the%20likelihood%20principle&f=false. Accessed March 20, 2012.

Besag, J., P.J. Green, D.M. Higdon, and K. Mengerson. 1995. Bayesian Computation and Stochastic Systems. *Statistical Science* 10:3-66.

Box, G., and N. Draper. 1987. *Empirical Model Building and Response Surfaces*. New York: Wiley.

Box, G.E.P., J.S. Hunter, and W.G. Hunter. 2005. *Statistics for Experimenters: Design Innovation, and Discovery, Volume 2*. New York: Wiley Online Library.

Brooks, H.E., and C.A. Doswell III. 1996. A Comparison of Measures-Oriented and Distributions-Oriented Approaches to Forecast Verification. *Weather Forecasting* 11:288-303.

Buser, C.M., H.R. Kunsch, D. Luth, M. Wild, and C. Schar. 2009. Bayesian Multi-Model Projection of Climate: Bias Assumptions and Interannual Variability. *Climate Dynamics* 33(6):849-868.

Christen, J.A., and C. Fox. 2005. Markov Chain Monte Carlo Using an Approximation. *Journal of Computational and Graphical Statistics* 14(4):795-810.

Cleveland, W.S. 1984. *Elements of Graphing Data*. Belmont, Calif.: Wadsworth.

Cook, R.D., and S. Weisberg. 1999. *Applied Regression Including Computing and Graphics*. New York: Wiley Online Library.

Cooley, D. 2009. Extreme Value Analysis and the Study of Climate Change. *Climatic Change* 97(1):77-83.

Dubois, D., H. Prade, and E.F. Harding. 1988. *Possibility Theory: An Approach to Computerized Processing of Uncertainty*. New York: Plenum Press.

Easterling, R.G. 2001. *Measuring the Predictive Capability of Computational Models: Principles and Methods, Issues and Illustrations*. SAND2001-0243. Albuquerque, N.Mex.: Sandia National Laboratories.

Efendiev, Y., A. Datta-Gupta, X. Ma, and B. Mallick. 2009. Efficient Sampling Techniques for Uncertainty Quantification. In *History Matching Using Nonlinear Error Models and Ensemble Level Upscaling Techniques*. Washington, D.C.: Water Resources Research and American Geophysical Union.

Evans, S.N., and P.B. Stark. 2002. Inverse Problems as Statistics. *Inverse Problems* 18:R55.

Evensen, G. 2009. *Data Assimilation: The Ensemble Kalman Filter*. New York: Springer Verlag.

Ferson, S., V. Kreinovich, L. Ginzburg, D.S. Myers, and K. Sentz. 2003. *Constructing Probability Boxes and Dempster-Shafer Structures*. Albuquerque, N.M.: Sandia National Laboratories.

Flath, H.P., L.C. Filcox, V. Akçelik, J. Hill, B. Van Bloeman Waanders, and O. Glattas. 2011. Fast Algorithms for Bayesian Uncertainty Quantification in Large-Scale Linear Inverse Problems Based on Low-Rank Partial Hessian Approximations. *SIAM Journal on Scientific Computing* 33(1):407-432.

Fuentes, M., and A.E. Raftery. 2004. Model Validation and Spatial Interpolation by Combining Observations with Outputs from Numerical Models via Bayesian Melding. *Journal of the American Statistical Association, Biometrics* 6:36-45.

Furnish, M.D., M.B Boslough, and G.T. Gray. 1995. Dynamical Properties Measurements for Asteroid, Comet and Meteorite Material Applicable to Impact Modeling and Mitigation Calculations. *International Journal of Impact Engineering* 17(3):53-59.

Galbally, D.K., K. Fidkowski, K. Willcox, and O. Ghattas. 2010. Nonlinear Model Reduction for Uncertainty Quantification in Large-Scale Inverse Problems. *International Journal for Numerical Methods in Engineering* 81:1581-1608.

Gelfand, A.E., and S.K. Ghosh. 1998. Model Choice: A Minimum Posterior Predictive Loss Approach. *Biometrica* 85(1):1-11.

Gelman, A., X.L. Meng, and H. Stern. 1996. Posterior Predictive Assessment of Model Fitness via Realized Discrepancies. *Statistica Sinica* 6:733-769.

Ghanem, R., and A. Doostan. 2006. On the Construction and Analysis of Stochastic Predictive Models: Characterization and Propagation of the Errors Associated with Limited Data. *Journal of Computational Physics* 217(1):63-81.

Gneiting, T., and A.E. Raftery. 2005. Weather Forecasting with Ensemble Methods. *Science* 310(5746):248-249.

Goldstein, M., and J.C. Rougier. 2004. Probabilistic Formulations for Transferring Inferences from Mathematical Models to Physical Systems. *SIAM Journal on Scientific Computing* 26(2):467-487.

Hastie, T., R. Tibshirani, and J.H. Friedman. 2009. *The Elements of Statistical Learning: Data Mining, Inference, and Prediction*. New York: Springer.

Higdon, D., M. Kennedy, J.C. Cavendish, J.A. Cafeo, and R.D. Ryne. 2005. Combining Field Data and Computer Simulations for Calibration and Prediction. *SIAM Journal on Scientific Computing* 26(2):448-466.

Higdon, D., J. Gattiker, B.Williams, and M. Rightley. 2008. Computer Model Calibration Using High-Dimensional Output. *Journal of the American Statistical Association* 103(482):570-583.

Hills, R., and T. Trucano. 2002. *Statistical Validation of Engineering and Scientific Models: A Maximum Likelihood Based Metric*. SAND2001-1789. Albuequerque, N. Mex.: Sandia National Laboratories.

Hills, R.G., K.J. Dowding, and L. Swiler. 2008. Thermal Challenge Problem: Summary. *Computer Methods in Applied Mechanics and Engineering* 197:2490-2495.

Hoeting, J.A., D. Madilgan, A.E. Raftery, and C.T. Volinsky. 1999. Bayesian Model Averaging: A Tutorial. *Statistical Science* 15:382-401.

Kaipio, J.P., and E. Somersalo. 2005. *Statistical and Computational Inverse Problems*. New York: Springer.

Kaipio, J.P., V. Kolehmainen, I. Somersalo, and M. Vauhkonen. 2000. Statistical Inversion and Monte Carlo Sampling Methods in Electrical Impedance Tomography. *Inverse Problems* 16:1487.

Kennedy, M.C., and A. O'Hagan. 2001. Bayesian Calibration of Computer Models. *Journal of the Royal Statistical Society: Series B (Statistical Methodology)* 63:425-464.

Kersting, A.B., D.W. Efurd, D.L. Finnegan, D.J. Rokop, D.K. Smith, and J.L.Thompson. 1999. Migration of Plutonium in Ground Water at the Nevada Test Site. *Nature* 397(6714):56-59.

Kirk, B., J. Peterson, R. Stogner, and G. Carey. 2006. A C++ Library for Parallel Adaptive Mesh Refinement/Coarsening Simulations. *Engineering with Computers* 22(3-4):237-254.

Klein, R., S. Doebling, F. Graziani, M. Pilch, and T. Trucano. 2006. *ASC Predictive Science Academic Alliance Program Verification and Validation Whitepaper*. UCRL-TR-220711. Livermore, Calif.: Lawrence Livermore National Laboratory.

Klir, G.J., and B. Yuan. 1995. *Fuzzy Sets and Fuzzy Logic*. Upper Saddle River, N.J.: Prentice Hall.

Knupp, P., and K. Salari. 2003. *Verification of Computer Codes in Computational Science and Engineering*. Boca Raton, Fla.: Chapman and Hall/CRC.

Knutti, R., R. Furer, C. Tebaldi, J. Cermak, and G.A. Mehl. 2010. Challenges in Combining Projections in Multiple Climate Models. *Journal of Climate* 23(10):2739-2758.

Kumamoto, H., and E.J. Henley. 1996. *Probabilistic Risk Assessment and Management for Engineers and Scientists*. New York: IEEE Press.

Lehmann, E.L., and J.P. Romano. 2005. *Testing Statistical Hypotheses*. New York: Springer.

Lieberman, C., K. Willcox, and O. Ghattas. 2010. Parameter and State Model Reduction for Large-Scale Statistical Inverse Problems. *SIAM Journal on Scientific Computing* 32:2523-2542.

Loeppky, J., D. Bingham, and W.J. Welch. 2011. Computer Model Calibration or Tuning in Practice. *Technometrics*. Submitted for publication.

Long, K., R. Kirty, and B. Van Bloemen Waanders. 2010. Unified Embedded Parallel Finite Element Computations via Software-Based Frechet Differentiation. *SIAM Journal on Scientific Computing* 32(6):3323-3351.

Lorenc, A.C. 2003. The Potential of the Ensemble Kalman Filter for NWP—A Comparison with 4D-Var. *Quarterly Journal of the Royal Meteorological Society* 129:3183-3203.

Lucas, L.J., H. Owhadi, and M. Ortiz. 2009. Rigorous Verification, Validation, Uncertainty Quantification and Certification Through Concentration-of-Measure Inequalities. *Computer Methods in Applied Mechanics and Engineering* 57(51-52):4591-4609.

Marzouk, Y.M., and H.N. Najm. 2009. Dimensionality Reduction and Polynomial Chaos Acceleration of Bayesian Inference in Inverse Problems. *Journal of Computational Physics* 228:1862-1902.

Meehl, G.A., C. Covey, T. Delworth, M. Latif, B. McAvaney, J.F.B. Mitchell, B. Stouffer, and K.E. Taylor. 2007. The WCRP CMIP3 Multimodel Dataset. *Bulletin of the American Meteorological Society* 88:1388-1394.

Mosleh, A., D.M. Rasmuson, F.M. Marshall, and U.S. Nuclear Regulatory Commission. 1998. *Guidelines on Modeling Common-Cause Failures in Probabilistic Risk Assessment.* Washington, D.C.: Safety Programs Division, Office for Analysis and Evaluation of Operational Data, U.S. Nuclear Regulatory Commission.

Naevdal, G., L. Johnsen, S. Aanonsen, and D. E. Vefring. 2005. Reservoir Monitoring and Continuous Model Updating Using Ensemble Kalman Filter. *Society of Petroleum Engineers Journal* 10(1):66-74.

NRC (National Research Council). 2007. *Models in Environmental Regulatory Decision Making,* Washington, D.C.: National Academies Press.

NSF (National Science Foundation). 2010. Minutes of the Advisory Committee Meeting. April 1-2, 2010. Available at http://www.nsf.gov/attachments /117978/public/MPSAC_April_1-2_2010_Minutes_Final.pdf. Accessed March 20, 2012.

Neuman, S.P. and J.W. Wierenga. 2003. *A Comprehensive Strategy of Hydrogeologic Modeling and Uncertainty Analysis for Nuclear Facilities and Sites.* Washington, D.C.: U.S. Nuclear Regulatory Commission.

Oakley, J.E., and A. O'Hagan. 2004. Probabilistic Sensitivity Analysis of Complex Models: A Bayesian Approach. *Journal of the Royal Statistical Society: Series B (Statistical Methodology)* 66(3):751-769.

Oberkampf, W.L., and C. Roy. 2010. *Verification and Validation in Scientific Computing.* Cambridge, U.K.: Cambridge University Press.

Oberkampf, W.L., and T.G. Trucano. 2000. Validation Methodology in Computational Fluid Dynamics. American Institute of Aeronautics and Astronautics, AIAA 200-2549, Fluids 2000 Conference, Denver, Colo.

Oberkampf, W.L., T.G. Trucano, and C. Hirsch. 2004. Verification, Validation, and Predictive Capability in Computational Engineering and Physics. *Applied Mechanical Reviews* 57:345.

Oden, J.T., and S. Prudhomme. 1998. A Technique for A Posteriori Error Estimation of h-p Approximations of the Stokes Equations. *Advances in Adaptive Computational Methods in Mechanics* 47:43-63.

Oliver, D.S., B.C. Luciane, and A.C. Reynolds. 1997. Markov Chain Monte Carlo Methods for Conditioning a Permeability Field to Pressure Data. *Mathematical Geology* 29:61-91.

Picard, R.R. 2005. Importance Sampling for Simulation of Markovian Physical Processes. *Technometrics* 47(2):202-211.

Prudhomme, S., and J.T. Oden. 1999. On Goal-Oriented Error Estimation for Elliptic Problems: Application to Pointwise Errors. *Computation Methods in Applied Mechanics and Engineering* 176:313-331.

Rabinovich, S. 1995. *Measurement Errors, Theory and Practice.* New York: The American Institute of Physics.

Raftery, A.E. 1996. Hypothesis Testing and Model Selection via Posterior Simulation. Pp. 163-168 in *Practical Markov Chain Monte Carlo.* London, U.K.: Chapman and Hall.

Roache, P. 1998. Verification and Validation in Computational Science and Engineering. Socorro, N.Mex.: Hermosa Publishers.

Robins, J.M., A. van der Vaart, and V. Ventura. 2000. Asymptotic Distribution of P Values in Composite Null Models. *Journal of the American Statistical Association* 95(452):1143-1156.

Rougier, J., M. Goldstein, and L. House. 2010. *Assessing Model Discrepancy Using a Multi-Model Ensemble.* University of Bristol Statistics Department Technical Report #08:17. Bristol, U.K.: University of Bristol.

Sain, S.R., R. Furrer, and N. Cressie. 2011. A Spatial Analysis of Multivariate Output from Regional Climate Models. *Annals of Applied Statistics* 5(1):150-175.

Scheffer, M., J. Bascompte, W.A. Brock, V. Brovkin, S.R. Carpenter, V. Dakos, H. Held, H.E.H. Van Nes, M. Rietkerk, and G. Sugihara. 2009. Early-Warning Signals for Critical Transitions. *Nature* 461(7260):53-59.

Shafer, G. 1976. *A Mathematical Theory of Evidence.* Princeton, N.J.: Princeton University Press.

Smith, R.L., C. Tebaldi, D. Nychka, and L.O. Mearns. 2010. Bayesian Modeling of Uncertainty in Ensembles of Climate Models. *Journal of the American Statistical Association* 104(485):97-116.

Steinberg, S., and P. Roache. 1985. Symbolic Manipulation and Computational Fluid Dynamics. *Journal of Computational Physics* 57(2):251-284.

Strouboules, F., I. Babuska, D.K. Dalta, K. Copps, and S.K. Gangarai. 2000. A Posteriori Estimation and Adaptive Control of the Error in the Quantity of Interest. Part 1: A Posterioric Estimations of the Error in the Von Mises Stress and the Stress Intensity Factor. *Computational Methods in Applied Mechanics and Engineering* 181:261-294.

Tarantola, A. 2005. *Inverse Problem Theory and Methods for Model Parameter Estimation.* Philadelphia, Pa.: SIAM.

Tebaldi, C., and R. Knutti. 2007. The Use of the Multi-Model Ensemble in Probabilistic Climate Projections. *Philosophical Transactions of the Royal Society, Series A* 365:2053-2075.

Tebaldi, C., R.L. Smith, D. Nychka, and L.O. Mearns. 2005. Quantifying Uncertainty in Projections of Regional Climate: A Bayesian Approach to the Analysis of Multimodel Ensembles. *Journal of Climate* 18:1524-1540.

Thornton, J. 2011. No Testing Allowed: Nuclear Stockpile Stewardship Is a Simulation Challenge. *Mechanical Engineering-CIME* 133(5):38-41.

Tonkin, M., and J. Doherty. 2009. Calibration-Constrained Monte Carlo Analysis of Highly Parameterized Models Using Subspace Techniques. *Water Resources Research* 45(12):w00b10.

Wan, E.A., and R. Van Der Merwe. 2000. The Unscented Kalman Filter for Nonlinear Estimation. Pp.153-158 in *Adaptive Systems for Signal Processing, Communications, and Control Symposium 2000.* AS-SPCC/IEEE, Lake Louise, Alta., Canada.

Wang, S., W. Chen, and K.L. Tsui. 2009. Bayesian Validation of Computer Models. *Technometrics* 51(4):439-451.

Welch, G., and G. Bishop. 1995. *An Introduction to the Kalman Filter.* Technical Report 95-041. Chapel Hill: University of North Carolina.

Wu, C.F.J., and M. Hamada. 2009. *Experiments: Planning, Analysis, and Optimization.* New York: Wiley.

Youden, W.J. 1961. Uncertainties in Calibration. *Precision Measurement and Calibration: Statistical Concepts and Procedures* 1:63.

Youden, W.J. 1972. Enduring Values. *Technometrics* 14(1)1-15.

6

Making Decisions

6.1 OVERVIEW

The ultimate goal of verification, validation, and uncertainty quantification (VVUQ) activities is to assist decision makers in reaching informed decisions about an intended application. As such, VVUQ activities are part of a larger group of decision-support tools that include modeling, simulation, and experimentation. The role of VVUQ could be as simple as providing uncertainty bounds or a worst-case analysis for a particular risk metric, or it could be as complex as using more rigorous methods (such as the design of validation experiments or other applications requiring optimization under uncertainty) to compare various options.

This chapter discusses decisions that have to be made during VVUQ activities and presents two examples describing how VVUQ activities enhance the eventual decision about the intended application. The incorporation of models and simulations within a complete decision-making system is a deep and complex question. The report *Models in Environmental Regulatory Decision Making* (NRC, 2007) provides a good discussion of this broader topic, which is beyond the scope of the current report. The types of decision discussed here can be grouped into two broad categories: (1) decisions that arise as part of the planning and conduct of the VVUQ activities themselves and (2) decisions made with the use of VVUQ results about an application at hand. Sections 6.4 and 6.5 present detailed examples of VVUQ applications.

6.2 DECISIONS WITHIN VVUQ ACTIVITIES

The nature of VVUQ activities depends fundamentally on how the results will be used in the eventual decision concerning an application. For example, an effort to obtain conservative bounds for a risk metric will be quite different from a study to obtain a comprehensive uncertainty quantification (UQ) analysis that informs a strategy for reducing uncertainty over time.

Finding: It is important that, before VVUQ activities begin, decision makers and practitioners of VVUQ discuss and arrive at an agreement on how the results of VVUQ analyses will be used.

As discussed throughout this report, VVUQ involves a large number of activities, each requiring many decisions. For example, verification studies deal with the implementation of numerical algorithms, encompassing the choice and allocation of resources to different types of algorithm testing (analytical solutions for simplified physics, the method of manufactured solutions, and so on), among other issues. Similarly, software quality assurance involves decisions about the different types and the extent of software testing, coverage analysis, and so on. Validation studies also involve many activities, ranging from the choice of input space, to the design and fielding of experiments, to the selection of emulators, to the analysis of output data, and so on. Again, there are many important choices to be made within each of these activities and many important decisions to be made on how to trade off resources and time among them.

The results of a UQ study can help to inform the decision maker on the relative priorities among a broad set of choices. These choices can be viewed as a large set of possible trade-offs through which the uncertainty is managed (an uncertainty management trade space). Components of this trade space include the following:

- Fundamental improvements to physics models,
- Improvements to the integrated simulation and modeling capability,
- The design and conduct of computer experiments,
- The design and conduct of relevant constraining physical experiments, and
- The engineering and design of the system to tolerate the predicted uncertainty.

The first four of these activities are typically (although not exclusively) considered as decisions *within* VVUQ. The final activity is typically considered to occur *after* the completion of the VVUQ study (treated in more detail in Section 6.3). All of these activities require resources, which include employing domain experts, accessing computational and experimental facilities, and influencing engineering design decisions. Decision makers must allocate resources throughout the VVUQ process, keeping in mind the goal of the study. For example, decision makers must weigh the relative benefit of investing in improvements to the fidelity of a given physics model against the benefit of conducting relevant physical experiments for calibration. Computing resources must be allocated across studies investigating detailed convergence, model fidelity, and completeness of UQ ensembles. Physical experiments must be selected from choices ranging from experiments involving components to fully integrated experiments. In industrial contexts, it is not uncommon for there to be a single budget for the entire process of modeling, simulator development, and UQ analysis—and the trade-offs are then even more critical. Ideally, the VVUQ framework helps to inform decisions on the relative impact of these activities and can be used to prioritize the allocation of resources.

Regardless of how carefully and efficiently the activities are carried out, difficult decisions will have to be made during the course of VVUQ. These decisions will have to withstand subsequent scrutiny and review by independent third parties.

Adequate documentation and transparency about the VVUQ process will facilitate peer review and provide archival information for future studies. It is important that peer reviewers be given access to all relevant information, data, and computational models (including codes, where appropriate) used in the VVUQ process.

Finding: It is important to include in any presentation of VVUQ results the assumptions as well as the sources of uncertainty that were considered. Appropriate documentation and transparency about the process and body of knowledge that were used to assess and quantify uncertainties in the relevant quantities of interest are also crucial for a complete understanding of the results of the VVUQ analysis.

6.3 DECISIONS BASED ON VVUQ INFORMATION

Ultimately, decision makers are faced with a set of choices, each one of which will have certain advantages and disadvantages. Within this framework, decision makers must make trade-offs based on the analyses and the probabilities of the various scenarios. For example, someone in environmental management may have to choose between two remediation strategies for cleaning up a contaminated site. The decision maker could choose an option for monitored natural attenuation—in other words, leaving the site as is but closely monitoring it to make sure that the contamination does not spread to high-risk areas. Or the decision maker could choose a more active, but also more costly, procedure that might clean up the site. The choice of option will be based on several underlying computational models, each with its own set of uncertainties that need to be compared against one another.

In the case of the Stockpile Stewardship Program (SSP) described in Section 6.4, similar types of decisions must be made. The SSP has developed its own framework, known as Quantification of Margins and Uncertainty (QMU), that produces a quantity known as the margin-to-uncertainty (M/U) ratio (Goodwin and Juzaitis, 2006). If the M/U ratio is "large," diligence is required, but the safety, security, and reliability of the weapon system are assured. If the M/U ratio approaches unity, decision makers are presented with a variety of options that involve trade-offs between two broad categories: increasing the margin or reducing the uncertainty. Each of these choices involves decisions that must take into account computational models and uncertainties as well as stochastic variables.

In many cases (and in the two examples referred to above), there is a close analogy to several areas of optimization that can play an important role in the mathematical foundation for decision making based on VVUQ. For example, the field of multiobjective optimization (Miettinen, 1999) is focused on the development of methods and algorithms for the solution of problems that involve multiple objectives that must be simultaneously minimized. This leads to approaches that can be used to trade off among different options. Stochastic optimization (Ermoliev, 1988; Heyman and Sobol, 2003) is another relevant area, in which some of the design parameters or constraints are described by random variables. The theory for these types of problems could be used to develop better bounds on the uncertainties associated with each decision. The simulation optimization field has other approaches. One alternative approach is that of robust optimization. Here, one seeks to find optimal solutions over a broad range of nonstochastic but uncertain input parameters (Ben-Tal and Nemirovski, 2002). In this case, a robust solution is one that remains "optimal" under the entire range given for the uncertain input parameters (Taguchi et al., 1987). These types of solutions are desirable if available, because decision makers can be assured that whatever option they choose, the consequences of uncertain input parameters will not generate large changes from the optimal solutions. All of these examples indicate that optimization will be a central component of the mathematical foundation for decision making under uncertainty.

A summary of the body of information that enables an assessment of the appropriateness of a model and its ability to predict the relevant quantities of interest (QOIs), as well as inclusion of the key assumptions used to make the prediction, is a necessary part of reporting model results. This information will allow decision makers to understand better the adequacy of the model as well as the key assumptions and data sources on which the reported prediction and uncertainty rely. In addition, the finding regarding transparency and documentation stated in Section 6.2 should be made available to decision makers and peer reviewers.

It is important to recognize that a UQ study will often be an ongoing effort, with decision making happening throughout the study, with respect to both the study itself and the external applications. The climate-modeling case study discussed in Section 2.10 in Chapter 2 is an example of such an ongoing effort—only limited UQ information is currently available for use in policy decisions. This example highlights the need for the development of decision-making platforms that can be based on only partial or very limited UQ information. It also highlights the need to identify situations for which more detailed UQ characterization would give additional clarity for decision making.

6.4 DECISION MAKING INFORMED BY VVUQ IN THE STOCKPILE STEWARDSHIP PROGRAM

When the moratorium on nuclear testing was begun in 1992, the Department of Energy (DOE) established alternative means for maintaining and assessing the nation's nuclear weapons stockpile. The SSP was created to "ensure the preservation of the core intellectual and technical competencies of the United States in nuclear weapons" (U.S. Congress, 1994). The SSP must assess, on an annual basis, the safety, security, and reliability of the nuclear weapons stockpile in the absence of nuclear testing. A key product of the annual assessment process is a report on the state of the stockpile, issued by the directors of the three national security laboratories—Lawrence Livermore National Laboratory, Los Alamos National Laboratory, and Sandia National Laboratories[1]—to the Secretary of Energy, the Secretary of Defense, and the Nuclear Weapons Council. With this report in hand, the Secretary of Energy, the Secretary of Defense, and the Commander of the U.S. Strategic Command each write a letter to

---

[1] Sandia National Laboratories is a multiprogram laboratory managed and operated by Sandia Corporation, a wholly owned subsidiary of Lockheed Martin Corporation, for the Department of Energy's National Nuclear Security Administration under contract DE-AC04-94AL85000.

the President of the United States providing their individual views on the health of the stockpile and on whether nuclear-explosive testing should resume. The SSP assessment thus provides the technical basis for the President's decision on whether or not to resume nuclear testing.

QMU, a component in the assessment framework, is a decision-support process. This case study illustrates the importance of UQ to reducing uncertainties in the QMU process and, thus, enhancing the decision-making process. By quantifying the largest sources of uncertainty, UQ, in this case, is essential to inform resource allocation for reducing error in inputs to the models. It provides a means of quantifying and then communicating the confidence in the performance and operation of a nuclear weapon. QMU provides a framework for systematically including the results of modeling and simulation, ongoing nonnuclear experiments, legacy nuclear testing, and informed judgment from nuclear weapons design physicists. A summary of the QMU methodology and current state of practice can be found in a study conducted by the National Research Council (NRC, 2009). The key concepts in QMU are quantification of the margin, the threshold for acceptable performance, and quantification of the uncertainty, representing imperfect knowledge of the physics and manufacturing of the weapon system. UQ is fundamental to the QMU framework.

QMU is the assessment methodology for focusing the SSP on risks to the stockpile and for delivering and communicating UQ-informed recommendations. As stewards of its nuclear deterrence capability, decision makers responsible for the nation's nuclear weapons stockpile have a range of options to consider if an issue arises:

- Do nothing (accepting the risks identified through QMU);
- Reduce the uncertainty in the relevant QOIs by employing theory, simulation, and/or experiments;
- Modify the weapon system;
- Engage the military to alter the characteristics of and/or requirements for the weapon system;
- Modify operations within the nuclear weapons complex and/or Department of Defense; or
- Cease certification of the weapon system.

This decision space has high consequences because of its impact on the nation's nuclear deterrence. The options under consideration may have an impact on the deterrence posture or require significant expenditures or both. Given the consequences of these decisions, the SSP uses QMU as a common language in all interactions (internally, in peer review, and with stakeholders), demonstrating the rational basis for decisions relating to the U.S. stockpile. QMU, being quantitative, provides transparency to the decision-making process.

The QMU framework, as it pertains to the SSP, *requires* highly trained design and computational physicists; it does not deliver a mathematical procedure that can be executed independent of informed judgment. Design judgment, informed by quantitative input, is at the heart of the SSP and is an essential part of the decision-making process. Judgment is based on technical rigor, tempered by experience, and demonstrated by performance. It requires quantitative inputs from simulation models with a finite domain of validity, experiments, and theory. In essence, judgment is knowing what questions to ask and an ability to draw conclusions from inputs, which are often limited and sometimes conflicting. Judgment is *not* simply assertion, nor is it independent of hard technical assessment.

Working with the QMU framework involves several key activities:

- Determining key performance metrics against which margins will be assessed;
- Establishing the verification and validation bases for simulation models;
- Performing uncertainty quantification; incorporating uncertainty owing to input data, manufacturing variations, and model-form uncertainty (where possible); and model calibration (where necessary);
- Establishing thresholds in performance metrics and quantifying margins and uncertainties; and
- Delivering a documentation basis including both the margin and the uncertainty and subjecting those results to internal and external peer review.

Each of these activities is identified in the present report, although perhaps with slightly different language. The key performance metrics are analogous to the quantities of interest identified in a UQ study; establishing thresholds in those metrics is a key contribution from experienced design staff; and VVUQ and documentation are important here as they are in other VVUQ activities. Ultimately, the product is summarized as a margin with an associated uncertainty in the key performance metrics.

Establishing thresholds, margins, and uncertainties must accommodate the use of calibrated models. Simulation tools employed in the SSP use the best physics modeling achieved to date. However, even with the vast computing resources available to the National Nuclear Security Administration, fully first-principles models for nuclear weapon performance are not practicable. Given that, the UQ methodology must help inform how calibrated models diverge as simulation models are extrapolated away from the calibration point. This concept, discussed above, is an essential aspect of the application of UQ within the SSP.

An increase in margin can be achieved by altering the required characteristics of the weapon system, modifying operations across the complex, or modifying the weapon system itself. Altering the required characteristics of the weapon system must be decided in partnership with the Department of Defense and may not be an available option. Modifying operations or the weapon system itself may require significant financial expenditures. Further, these modifications may move the system away from the established calibration basis, an action that, by definition, introduces additional uncertainty. The increase in margin must compensate for this increase in uncertainty in order to deliver a net increase in the M/U ratio.

Reducing uncertainties can be achieved through the application of a breadth of capabilities available to the SSP. A UQ-informed recommendation is a key component in this decision-making process. QMU quantifies uncertainties compared to the assessed margins, focusing on margins that need to be increased or uncertainties that need to be reduced or better quantified. A main-effects analysis, which can quantify the largest sources of uncertainty, is essential in order to inform resource allocation for reducing input errors. Improvements may be obtained by improving the theory for the relevant physics model, which will constrain the model-form error and possibly eliminate the need for calibration. Improvements can also be achieved by conducting an experimental campaign to improve the calibration basis for simulation models, which constrains the error introduced by the calibration process. Similarly, experiments may be performed to expand the domain of validity of the simulation models, which could reduce prediction uncertainty for QOI or other key metrics. Finally, improvements can be achieved by reducing the errors associated with the model inputs, through improved input characterization, improved experimental data, or improved theory. Importantly, UQ-informed decisions help identify when, in a particular focus area, "enough is enough" and continued investment is unlikely to improve the overall confidence in system performance.

QMU provides a credible, quantifiable, and scientifically defensible basis for making programmatic decisions that impact planning, prioritizing, integrating, and communicating across various elements of the SSP. A strategy to fulfill the nation's stockpile decision-making responsibilities must provide a balance of capabilities—informed by QMU—across experimentation, physics models, simulation tools, theory, and analysis methods. UQ-informed simulation capabilities enable the use of high-fidelity design studies to provide U.S. policy makers with options for the future, including options that affect the stockpile, the weapons manufacturing complex, and experimental facilities.

6.5 DECISION MAKING INFORMED BY VVUQ AT THE NEVADA NATIONAL SECURITY SITE

6.5.1 Background

The context of this case study involving the Nevada National Security Site is Yucca Flat, Nevada Test Site (Figure 6.1), where a total of 659 underground nuclear tests were conducted between 1951 and 1992 (Fenelon, 2005). Most of the deep, large tests at Yucca Flat were conducted beneath the water table in complex layers of volcanic rocks dissected by numerous faults. Numerical models are being developed to predict the migration of radionuclides in the groundwater system away from the test cavities over the next 1,000 years. Of particular concern is a regionally extensive carbonate aquifer just below the volcanic layers. Predictions such as the size and extent of the plume as it evolves over time, the total volume of water exceeding the Safe Water Drinking Act standards, and the mass flux of radionuclides to the lower aquifer must be placed in a probabilistic framework. For the purpose of the study discussed in this report, which does not yet consider contaminant transport, the QOI is the increase in water volume flowing from the study site into the lower aquifer over the next 1,000 years. This case study also

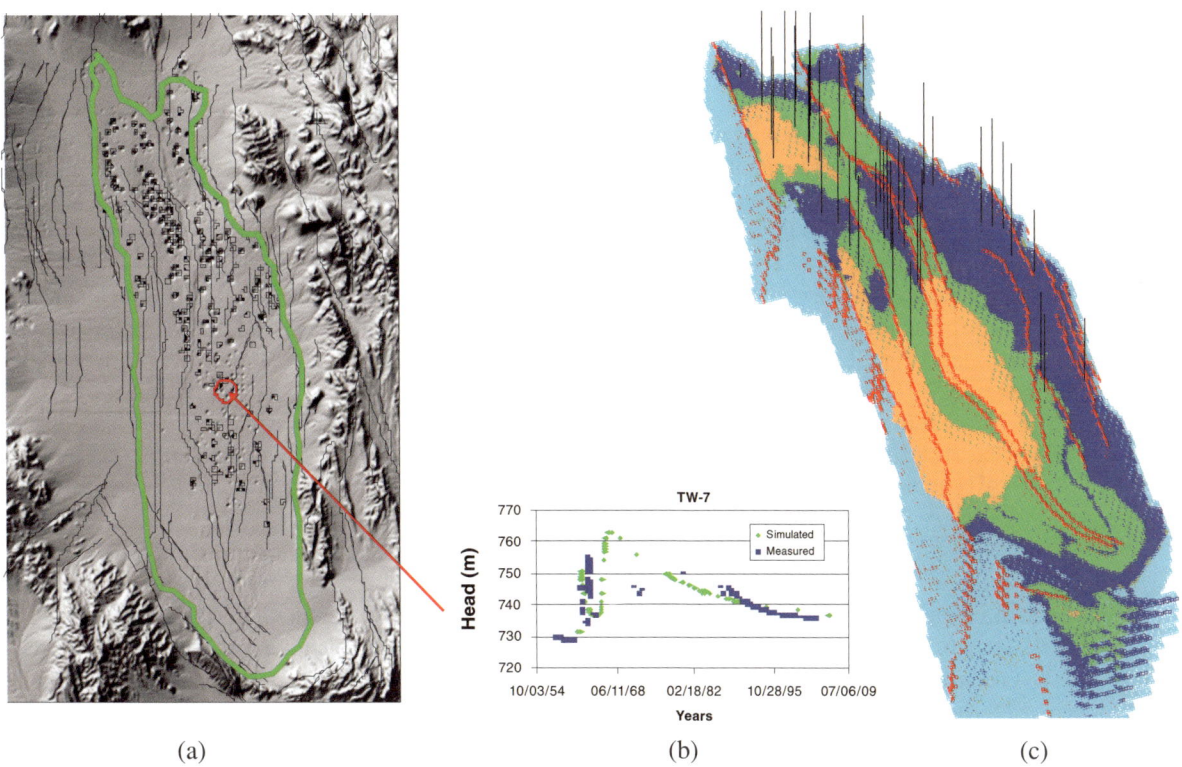

FIGURE 6.1 Yucca Flat, Nevada Test Site. (a) Study site locations of nuclear tests are denoted by black circles. The green outline depicts the region being modeled. The red circle indicates the location of the test well whose data are plotted in (c). (b) Hydraulic area level data over time for the test well identified by the red circle in (a). (c) Computation model grid for the high-fidelity model includes aquifers (light and dark blue), aquitards (orange and green), and faults (red) lines. The black lines denote well locations. SOURCE: Keating et al. (2010).

illustrated the use of UQ to quantify uncertainty in the decision-making process. In this case, however, the QOI under study accumulates over 1000 years and, so, cannot be observed. The DOE and the Nevada Division of Environmental Protection will be making key decisions concerning remediation, monitoring, and future hydrogeologic data collection on the basis of the results of this larger investigation. Although verification and validation play crucial roles in this larger effort, this case study focuses on the UQ aspects, showing how results are obtained and how they affect communication and decision making.

This case study has two main phases: (1) a calibration phase, in which available measurements are used to constrain uncertain parameters in the model, and (2) a prediction phase, in which prediction uncertainty is estimated for the QOI.

6.5.2 The Physical System

There are many important processes to be considered in this example, some focusing heavily on the source term (underground nuclear tests) and others focusing on flow and transport processes in the groundwater. In the former category are processes that cause drastic alteration in rock properties and pore water pressures near the working point and processes that distribute radionuclides in the cavity and beyond. All of these processes occurred in the area surrounding each test site within the first few seconds after a blast. In the latter category were processes such as advection, diffusion, dispersion, mass transfer between fractures and rock matrix, ground-surface subsidence due to long-term depressurization of the aquifer, sorption to mineral surfaces, and radioactive decay.

6.5.3 Computational Modeling of the Physical System

The study models the system using two iterative-coupled codes. The first is a phenomenological testing-effects model that simulates the instantaneous rock and fluid pressure changes that are due to underground testing. The second is a finite-volume heat- and mass- transfer code (Zyvoloski et al., 1997) used to simulate the transient flow of groundwater and the transport of radionuclides. The coupled model requires approximately 7 hours to run on a 3.6- GHz processor on a finely resolved numerical mesh (Figure 6.1(c)). Only the testing-effects and groundwater flow (not the radionuclide transport) portion of the simulations are considered in this report. An understanding of how testing effects groundwater flow is key to addressing the eventual question of radionuclide transport.

Several hundred parameters are used in this model. Many of these are related to the (possibly) unique characteristics of each underground nuclear test. Others are related to the permeability, porosity, and storage characteristics of the various rock layers and fault zones. Many (if not most) of the parameters are associated with essentially irreducible uncertainty. However, it is important to constrain as many parameters as possible. Transient hydraulic head data, basically measuring pressure at the monitoring wells (black lines in Figure 6.1(c)), were collected during the period of testing at approximately 60 wells; these data were used for the parameter estimation.

6.5.4 Parameter Estimation

The model-calibration process uses the hydraulic head measurements, capturing pressure at the monitoring wells, collected at various times for the approximately 60 wells. Measurements at various times are shown for one of the wells in Figure 6.1(b). The goal is to use these calibration data to constrain parameter uncertainty, which can then be used to produce uncertainties for the prediction of the QOI. This prediction with its associated uncertainty can then be used to help make decisions regarding additional monitoring or mitigation.

It is important that uncertainty be adequately captured in applications such as this one, in which no direct measurements of the QOI are available. This UQ analysis was designed with the goal that it not only would produce a reasonably well-calibrated model but at the same time would establish the framework for the following uncertainty-assessment phase for the QOI. These two goals can sometimes be in conflict in the following sense: traditional calibration methods fail when parameter dimensionality is large compared to data availability (the ill-posed inverse problem), a frequent occurrence in hydrogeologic inverse problems. A common strategy for dealing with this ill-posedness is to "fix" most model parameters at a nominal setting, allowing only those that are sensitive to the calibration data to vary. Unfortunately, although this parsimony-based strategy can be successful in producing a set of parameters that are close to the measurements, it can vastly underrepresent parameter uncertainty, producing inappropriately certain estimates of the QOI (Hunt et al., 2007). Here the QOI is the increase in water volume flowing from the study area to the lower aquifer over the next 1,000 years.

A Bayesian formulation of the inverse problem was used to describe the uncertainty in the 200+ parameters induced by conditioning on the calibration data. This requires the specification of a distribution and range for each of the input parameters, as well as the specification of a likelihood for the hydraulic head measurements given the model parameters. An approach called null-space Monte Carlo (Tonkin and Doherty, 2009) was used to produce samples from the resulting posterior distribution, which is available in the PEST[2] software suite (Doherty, 2009). This approach uses derivative-based searching as well as Monte Carlo sampling of the posterior density. A key feature of this analysis is that the large number of parameters not constrained by calibration data can be freely varied in the uncertainty analysis, producing a wider range of outcomes for the QOI.

This forward model is highly nonlinear and computationally demanding, which makes it difficult to tune and assess UQ methods for this application. To facilitate the development and testing of a suitable parameter-estimation and uncertainty-analysis strategy for this application, a fast-running reduced model was constructed. This reduced model (Keating et al., 2010) has a number of parameters similar to that of the process model and could be calibrated against the same data set described above. The null-space Monte Carlo approach, after tuning and testing using this reduced model, is ported over to and used for the calibration of the high-fidelity CPU-intensive process model.

---

[2] See pesthomepage.org. Accessed September 7, 2011.

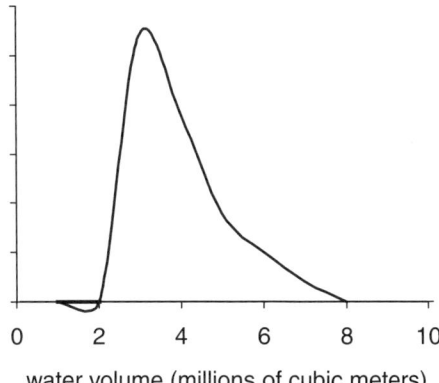

FIGURE 6.2 An ensemble of predictions of the quantity of interest—the additional water volume expected to move to the lower aquifer owing to nuclear testing effects, integrated over the study region and over the time span of 1,000 years at the Yucca Flat, Nevada Test Site. Even the highest values of this prediction are less than 1 percent of the total volume of water moving into the lower aquifer over this 1,000-year time span. SOURCE Keating et al. (2010).

6.5.5 Making (Extrapolative) Predictions and Describing Uncertainty

The posterior sample of parameter settings given the calibration data (which were found in the previous section) was then propagated forward through the computationally intensive process model, generating an ensemble of predictions for the QOI—the additional water volume attributable to the effects of nuclear testing, integrated over the study region (outlined in Figure 6.1(a)), and integrated over the span of 1,000 years. A probability density function for the QOI estimated from this ensemble is shown in Figure 6.2. The size of the ensemble was fairly small due to computational resource constraints.

To help assess the reliability of the prediction for the QOI, an ensemble of predictions was generated for a quantity that had not been used in model calibration—ground-surface subsidence (Keating et al., 2010). It was found that for nearly 90 percent of the model domain, measured subsidence fell within the bounds of the ensemble of predictions, giving some indication that this model can extrapolate from well-head measurements to other important model outputs (Keating et al., 2010).

Additionally, a number of discrete alternative conceptual models were created, primarily addressing key issues concerning the coupling of testing phenomenology and rock mechanics. These are essentially irreducible uncertainties. The alternative models were equally well calibrated to measured field data and could be considered to be equally plausible. The range of predictions generated using these discrete models was low, however, providing confidence in the robustness of results based on any single model.

6.5.6 Reporting Results to Decision Makers and Stakeholders

During the model development, analysis, and predictive phases, frequent briefings with the stakeholders (DOE and Nevada Division of Environment Protection) were held. The briefings provided the opportunity for stakeholders to give input and feedback and to gain confidence with the transparency of the process. It was particularly important to ensure that any and every credible conceptual model be included in the uncertainty analysis.

6.6 SUMMARY

As detailed in this chapter, a decision can be made regarding the allocation of resources during a VVUQ study, as well as after the VVUQ analyses are completed, the results serving as a key input to the decision-making process. Clearly, the information that goes into such a decision is likely to be both qualitative and quantitative. Also, decisions within the VVUQ process can be made to improve both qualitative and quantitative aspects of the information, just as might be done for designing validation experiments.

In cases where quantitative results are needed, optimality criteria will likely involve a mathematical summary of information or prediction uncertainty. The computations required to carry out such optimization searches are typically quite demanding, making the discussion in Chapter 4 on emulation and reduced-order models relevant here. Also, even when qualitative information is desired, it is often obtained through a quantitative analysis. This was the case in the case study presented in Section 6.5, in which quantitative information about ground surface subsistence was used to produce qualitative information regarding the prediction uncertainty for the QOI.

Finding: High-consequence decisions have been and continue to be informed by UQ assessments.

6.7 REFERENCES

Ben-Tal, A., and A. Nemirovski. 2002. Robust Optimization—Methodology and Applications. *Mathematical Programming, Series B 92*, pp. 453-480.

Doherty, J. 2009. *Addendum to the PEST Manual*. Corinda, Australia: Watermark Numerical Computing.

Ermoliev, Y. 1988. *Nonlinear Multiobjective Optimization*. New York: Springer.

Fenelon, J.M. 2005. *Analysis of Ground-Water Levels and Associated Trends in Yucca Flat, Nevada Test Site, Nye County, Nevada, 1951-2003*. U.S. Geological Survey Scientific Investigations Report 2005-5175. Washington, D.C.: U.S. Department of the Interior.

Goodwin, B.T., and R.J. Juzaitis. 2006. *National Certification Methodology for the Nuclear Weapon Stockpile*. UCRL-TR-223486. Available at http://www.osti.gov/bridge/product.biblio.jsp?_id=929177. Accessed March 19, 2011.

Heyman, D.P., and M.J. Sobel. 2003. *Stochastic Models in Operations Research, Vol. II: Stochastic Optimization*. Mineola, N.Y.: Dover Publications.

Hunt, R.J., J. Doherty, and M.J. Tonkin. 2007. Are Models Too Simple? Arguments for Increased Parameterization. *Ground Water* 45(3):254-262.

Keating, E.H., J. Doherty, J.A. Vrugt, and Q. Kang. 2010. Optimization and Uncertainty Assessment of Strongly Nonlinear Groundwater Models with High Parameter Dimensionality. *Water Resources Research* 46(10):W10517.

Miettinen, K. 1999. *Nonlinear Multiobjective Optimization*. New York: Springer.

NRC (National Research Council). 2007. *Models in Environmental Regulatory Decision Making*. Washington, D.C.: National Academies Press.

NRC. 2009. *Evaluation of Quantification of Margins and Uncertainties Methodology for Assessing and Certifying the Reliability of the Nuclear Stockpile*. Washington, D.C.: The National Academies Press.

Oldenburg, C.M., B.M. Freifeld, K. Pruess, L. Pan, S. Finsterle, and G.J. Moridis. 2011. Numerical Simulations of the Macondo Well Blowout Reveal Strong Control of Oil Flow by Reservoir Permeability and Exsolution of Gas. *Proceedings of the National Academy of Sciences*: July.

Taguchi, G., L.W. Tung, and D. Clausing. 1987. *System of Experimental Design: Engineering Methods to Optimize Quality and Minimize Costs*. New York: Unipub.

Tonkin, M., and J. Doherty. 2009. Calibration-Constrained Monte Carlo Analysis of Highly Parameterized Models Using Subspace Techniques. *Water Resources Research* 45(12):w00b10.

U.S. Congress. 1994. Section 3138, National Defense Authorization Act for the Year 1994. Public Law 103-160; 42 U.S.C. 2121 Note.

Zyvoloski, G.A., B.A. Robinson, Z.V. Dash, and L.L. Trease. 1997. *User's Manual for the FEHM Application—A Finite-Element Heat- and Mass-Transfer Code*. Los Alamos, N.Mex.: Los Alamos National Laboratory.

7

Next Steps in Practice, Research, and Education for Verification, Validation, and Uncertainty Quantification

The role of verification, validation, and uncertainty quantification (VVUQ) in computational science and engineering has increased significantly in recent years. As high-quality computational modeling becomes available in more application areas, the role played by VVUQ will continue to grow. Previous chapters have addressed VVUQ as it has evolved to date in the computational modeling of complex physical systems. In this chapter, the committee discusses next steps in the evolution of VVUQ. This summary of its responses to the statement of task, includes the committee's identification of principles and current best practices and its recommendations for VVUQ research and development, as well as recommendations for educational changes.

7.1 VVUQ PRINCIPLES AND BEST PRACTICES

As was noted in Chapter 1, the committee has confined its considerations of principles and best practices to the mathematical science aspects of VVUQ. The principles and best practices presented here are, loosely, restricted to those aspects and do not emphasize nonmathematical issues of physical science, communication of results, and so forth. Historically, methodologies for VVUQ have evolved separately in different application areas and fields. As a result, different application areas can have different approaches. A number of recent workshops and conferences have assembled researchers from varied application areas and perspectives, aiming for a cross-fertilization of ideas and a better understanding of the connections, commonalities, and differences among the varied VVUQ practices. As time passes, the relationships among the various practices developed in different settings will become clearer, as will the understanding of best practices for different kinds of applications. However, it is premature to try to identify a single set of methods or algorithms that are the best tools to accomplish the best practices identified below. Today, it appears that some methods and algorithms are better for some applications and others are better for other applications. Therefore, the committee identifies principles and best practices but stops short of prescribing implementation methodologies.

This section begins with some overarching remarks, moves to principles and practices for verification, and then addresses principles and practices for validation and prediction. As previous chapters have emphasized, VVUQ analyses are not well defined unless the quantities of interest (QOIs) are well defined. Defining the QOIs from the start allows a VVUQ process to produce more meaningful results than will be produced if the focus is on the "solution" in general. For example, suppose a given model accurately captures the average, or large-scale, features of a physical system but not the small-scale features. If only large-scale features are important in the given application, the appropriately defined QOI should be sensitive to large-scale but not small-scale behavior. In this case the VVUQ analysis may find that the model is sufficiently accurate (e.g., uncertainties in the predicted QOI are sufficiently small) to provide actionable information. However, if small-scale details are important, the QOI should be defined accordingly, and the VVUQ analysis (of the same model applied to the same physical system) may find that the model is too inaccurate to be of value.

Leveraging work from previous VVUQ analyses should be done with caution. Since VVUQ results are specific to particular QOIs in particular settings, transferring results to new QOIs and settings can be difficult to justify. However, one can consider applying VVUQ to a model over a broad set of conditions and QOIs if physical data are available to support such wide-ranging assessments of model accuracy and there is a firm theoretical understanding of the physical phenomena being modeled. It can be argued that an example of such a situation is the Monte Carlo N-Particle transport code,[1] a particle-transport code that incorporates a large body of knowledge and has been tested against measurements derived from thousands of experiments spanning many particle types and a broad range of conditions.

Within the VVUQ enterprise, the level of rigor employed should be commensurate with the importance and needs of the application and decision context. Some applications involve high-consequence decisions and therefore require a substantial VVUQ effort; others do not.

7.1.1 Verification Principles and Best Practices

Here the committee summarizes key verification principles, along with best practices associated with each principle. Chapter 3 provides more detail.

- Principle: Solution verification is well defined only in terms of specified quantities of interest, which are usually functionals of the full computed solution.
    —Best practice: Clearly define the QOIs for a given VVUQ analysis, including the solution verification task. Different QOIs will be affected differently by numerical errors.
    —Best practice: Ensure that solution verification encompasses the full range of inputs that will be employed during UQ assessments.
- Principle: The efficiency and effectiveness of code and solution verification can often be enhanced by exploiting the hierarchical composition of codes and mathematical models, with verification performed first on the lowest-level building blocks and then on successively more complex levels.
    —Best practice: Identify hierarchies in computational and mathematical models and exploit them for code and solution verification. It is often worthwhile to design the code with this approach in mind.
    —Best practice: Include in the test suite problems that test all levels in the hierarchy.
- Principle: Verification is most effective when performed on software developed under appropriate software quality practices.
    —Best practice: Use software configuration management and regression testing and strive to understand the degree of code coverage attained by the regression suite.
    —Best practice: Understand that code-to-code comparisons can be helpful, especially for finding errors in the early stages of development but that in general they do not by themselves constitute sufficient code or solution verification.
    —Best practice: Compare against analytic solutions, including those created by the method of manufactured solutions—a technique that is helpful in the verification process.
- Principle: The goal of solution verification is to estimate, and control if possible, the error in each QOI *for the problem at hand*. (Ultimately, of course, one would want to use UQ to facilitate the making of decisions in the face of uncertainty. So it is desirable for UQ to be tailored in a way to help identify ways to reduce uncertainty, bound it, or bypass the problem, all in the context of the decision at hand. The use of VVUQ for uncertainty management is discussed in Section 6.2, "Decisions Within VVUQ Activities".)

---

[1] See mcnp-green.lanl.gov. Accessed September 7, 2011.

—Best practice: When possible in solution verification, use goal-oriented a posteriori error estimates, which give numerical error estimates for specified QOIs. In the ideal case the fidelity of the simulation is chosen so that the estimated errors are small compared to the uncertainties arising from other sources.

—Best practice: If goal-oriented a posteriori error estimates are not available, try to perform self-convergence studies (in which QOIs are computed at different levels of refinement) on the problem at hand, which can provide helpful estimates of numerical error.

—Remark: In the absence of a posteriori or self-convergence results, the next best option may be to estimate numerical error in a given QOI in the problem at hand based on detailed assessments of numerical error in a similar QOI in a relevant reference problem. However, it is challenging to define reference problems that permit detailed assessments but are demonstrably relevant to the problem at hand. It can be risky to assume that numerical errors in the reference problem are representative of numerical errors in the problem at hand.

7.1.2 Validation and Prediction Principles and Best Practices

Although the questions involving solution verification are firmly grounded in mathematical and computational science, the questions that arise in validation and prediction require statistical and subject-matter (physics, chemistry, materials, etc.) expertise as well. They also require choices that involve judgment, for example in determining the relevance of validation studies to the prediction of a QOI in the problem at hand. This necessary application of judgment warrants a brief discussion here. The concept of a domain of applicability—a region of a domain space in which a validation assessment is judged to apply—is helpful in determining the relevance of a validation assessment to the prediction of a QOI in a given problem at hand. This concept can include features, or descriptors, that characterize the problem space (such as ball density, radius, and drop height in the ball-drop example) as forming axes that define a mathematical space. Each problem or experiment is associated with a point in the space; thus, the problems included in the validation assessment map to a collection of points in the domain space. The problem at hand maps to another point in the space. One can imagine basing a determination of relevance on the location of a particular problem point relative to the locations of the other points. For example, if the new point is surrounded by validation-problem points, the validation study might be judged to have high relevance.

This is an appealing notion, but any attempt to apply it with mathematical rigor must address significant complicating truths. If important features are omitted from the set that is chosen to form axes of the space, then two problems may look similar when they actually differ in important ways. This is illustrated by the "ball texture" in the ball-drop example in Section 1.6. However, if all potentially important features are included, the dimension of the space may become intractably large. In the ball-drop example, potential additional features could include ambient temperature, ambient pressure, ambient humidity, wind conditions, ball skin materials, ball interior structure, initial rotation applied to the ball as it is dropped, ball elasticity, ball coefficient of thermal expansion, and so on. Including a large set of features would help guard against the omission of those that may be important. However, this creates a high-dimensional domain space, which forces essentially any new problem to be "outside" the region enclosed by previous problems. This makes every prediction appear to be "extrapolative." In oversimplified terms: if the domain space is low-dimensional, then subject-matter judgment is required to assess the impact of features that are not included, but if the domain space is high-dimensional, subject-matter expertise is required to assess the relevance of previous experience to an extrapolative prediction. Either way, subject-matter expertise must inform a judgment.

This discussion is not intended to attack the concept of a domain of applicability or to downplay its utility. Rather, it is intended to illustrate that mathematics alone cannot determine the relevance of past experience to the problem at hand but that judgment informed by subject-matter expertise is a necessary ingredient in making this determination.

In spite of variations in validation and prediction practices across fields, the inherent role of expertise and judgment, and the rapid evolution of improved methodologies, some general principles and best practices in validation and prediction have emerged that the committee believes will stand the test of time. They are summarized below. Chapter 5 provides more detail.

- Principle: A validation assessment is well defined only in terms of specified QOIs.
  —Best practice: Early in the validation process, specify the QOIs that will be addressed.
- Principle: A validation assessment provides direct information about model accuracy only in the domain of applicability that is "covered" by the physical observations employed in the assessment.
  —Best practice: When quantifying or bounding model error for a QOI in the problem at hand, systematically assess the relevance of supporting validation assessments (which were based on data from different problems, often with different QOIs). Subject-matter expertise should inform this assessment of relevance (as discussed above and in Chapter 5).
  —Best practice: If possible, use a broad range of physical observation sources so that the accuracy of a model can be checked under different conditions and at multiple levels of integration.
  —Best practice: Use "holdout tests" to test validation and prediction methodologies. In such a test some validation data are withheld from the validation process, the prediction machinery is employed to "predict" the withheld QOIs, with quantified uncertainties, and finally the predictions are compared to the withheld data.
  —Best practice: If the desired QOI was not observed for the physical systems used in the validation process, compare sensitivities of the available physical observations with those of the QOI.
  —Best practice: Consider multiple metrics for comparing model outputs against physical observations.
- Principle: The efficiency and effectiveness of a validation assessment are often improved by exploiting the hierarchical composition of computational and mathematical models, with assessments beginning on the lowest-level building blocks and proceeding to successively more complex levels.
  —Best practice: Identify hierarchies in computational and mathematical models, seek measured data that facilitate hierarchical validation assessments, and exploit the hierarchical composition to the extent possible.
  —Best practice: If possible, use physical observations, especially at more basic levels of the hierarchy, to constrain uncertainties in model inputs and parameters.
- Principle: The uncertainty in the prediction of a physical QOI must be aggregated from uncertainties and errors introduced by many sources, including: discrepancies in the mathematical model, numerical and code errors in the computational model, and uncertainties in model inputs and parameters.
  —Best practice: Document assumptions that go into the assessment of uncertainty in the predicted QOI, and also document any omitted factors. Record the justification for each assumption and omission.
  —Best practice: Assess the sensitivity of the predicted QOI and its associated uncertainties to each important source of uncertainty as well as to key assumptions and omissions.
  —Best practice: Document key judgments—including those regarding the relevance of validation studies to the problem at hand—and assess the sensitivity of predicted QOI and its associated uncertainties to reasonable variations in these judgments.
- Principle: Validation assessments must take into account the uncertainties and errors in physical observations (measured data).
  —Best practice: Identify all important sources of uncertainty/error in validation data—including instrument calibration, uncontrolled variation in initial conditions, variability in measurement setup, and so on—and quantify the impact of each.
  —Best practice: If possible, use replications to help estimate variability and measurement uncertainty.
  —Remark: Assessing measurement uncertainties can be difficult when the "measured" quantity is actually the product of an auxiliary inverse problem—that is, when the quantity is not measured directly but is inferred from other measured quantities.

7.2 PRINCIPLES AND BEST PRACTICES IN RELATED AREAS

7.2.1 Transparency and Reporting

In the presentation of VVUQ results to stakeholders, including decision makers who may not be familiar with the analyses, it is important to state clearly the key underlying assumptions along with their potential impact on the predicted QOIs, their uncertainties, and other key outcomes. In particular, UQ analyses should state which uncertainties are accounted for and which are not and should give some assessment of the impact of those uncertainties not accounted for. It is also important that the presentation discuss a triage of assumptions, assessing which have the potential to alter the outcomes and assessing the sensitivity of key outcomes to these alternative assumptions. A good example of detailing model inadequacies that might affect overall assessment of anthropogenic impact on climate change is given in Chapter 8 of an Intergovernmental Panel on Climate Change report (Randall et al., 2007).

The use of plain language, suitable for the application at hand, is most effective for presentations. The use of terminology that has specific meanings in mathematics, statistics, or VVUQ can often lead to misconceptions or misunderstandings. As Oreskes et al. (1994) point out, words such as "verification" and "validation" carry common meanings that can be inappropriately attached to computational-model assessment. It is also important not to confuse the mathematical, computational, and subject-matter science that went into building a large-scale computational model with the VVUQ effort that assesses the appropriateness and accuracy of the model-based predictions.

Holdout tests provide a direct demonstration of a model's ability to predict under new conditions and can be an effective tool for communicating certain VVUQ concepts and results. Holdout tests use the model to predict experimental or observational outcomes that were not used in the model calibration process. Once a computational model has been calibrated with a particular set of physical measurements, the holdout test allows one to see how the model predicts the system behavior in a new setting. Of course, assessing the degree of extrapolation in a given holdout test is still an open question, as the committee has discussed above.

7.2.2 Decision Making

Decision makers must have key information from the VVUQ process that is summarized and clearly communicated. This key information includes summaries of the body of knowledge behind the choice of models, evidence from the verification process, sensitivities of the calculated QOIs to uncertainties in key parameters, quantification (from validation studies) of the model's ability to match relevant measured data, assessment of modeling challenges in the prediction problem relative to those in the validation problems, key assumptions behind the predictions and quantified uncertainties, sources of uncertainty that were neglected, and so on. If this information is summarized and communicated properly, results from the VVUQ process can play a unique and significant role in the efficient allocation of resources, management of the overall uncertainty budget, and generation of the soundest possible basis for high-consequence decisions in the presence of uncertainties.

The results of VVUQ analyses can also be used to make decisions regarding how to allocate resources for future VVUQ activities—computing hardware acquisition, experimental campaigns, model improvement efforts, and other efforts—to improve prediction accuracy or to improve confidence in model-based predictions. This decision task is made more difficult by the often high cost of employing available computational models and the inability of models to perfectly represent reality. A realistic assessment of model inadequacies/discrepancies is important for resource allocation because models can inform only about processes represented in the models. If better understanding of current model inadequacies is key to improving predictions, then additional validation data will likely be required. Hence approaches for resource allocation will necessarily require some form of qualitative assessments or judgment. Given the complexity of VVUQ activities, a carefully structured planning process can help to ensure that resources are used efficiently and that significant factors are addressed.

7.2.3 Software, Tools, and Repositories

Practitioners in VVUQ currently have available to assist them a limited set of software and repositories (for data, examples, and code). This is particularly true for the developing field of uncertainty quantification. A number of application-specific software projects have been developed—Dakota,[2] for engineering applications, and PEST,[3]

---

[2] See Dakota.sandia.gov. Accessed September 7, 2011.
[3] See pesthomepage.org. Accessed September 7, 2011.

for environmental applications, are two notable examples. There is also software available to carry out specific computations involved in the VVUQ process (e.g., sensitivity analysis, response-surface modeling, logical-error checking for code verification, and so on.). A recently launched Department of Energy (DOE) effort is focused on developing software tools for UQ in the high-performance computing environment.

Such software as that described above can benefit practitioners and users. The more established efforts have documentation and a user community to help with their use. Although the learning curve is steep, and the framework and tools imposed by a particular software package may not be ideal for the application at hand, many of the utilities in current and developing software would be of use in many VVUQ efforts. Separate, usable libraries of functions and utilities could be used internally in other software efforts, which would make them more useful.

Nearly all of the available software treats the computational model as a black box that produces outputs for a given input setting. Such an approach has obvious advantages for general use—it requires no changes to existing computational models—but will be difficult to adapt to newer, intrusive approaches for UQ.

The VVUQ field would benefit from a collection of testbed examples that demonstrate software and VVUQ methods, provide examples of UQ analyses, and so on. Such a repository, perhaps managed by the Society for Industrial and Applied Mathematics, the American Statistical Association, or some other professional entity with a stake in VVUQ, would allow for the comparison and assessment of different methods and approaches so that practitioners could determine the most appropriate method(s) for their particular application. Such a repository would also help foster an understanding of the similarities in and differences among the various VVUQ approaches that have been developed in separate application areas.

7.3 RESEARCH FOR IMPROVED MATHEMATICAL FOUNDATIONS

This section discusses research directions that could improve the mathematical foundations of the VVUQ process. In the area of solution verification there is a need for methods that can accurately estimate numerical error in the computation of the problem at hand for mathematical models that are more complex than linear elliptic partial differential equations. In the area of validation and prediction, research needs are driven largely by (1) the computational burden presented by large-scale computational models, (2) the need to combine multiple sources of information, and (3) the challenges associated with assessing the quality of model-based predictions. In the area of uncertainty quantification, there is a need for improved methods for handling large numbers of uncertain inputs (the famous "curse of dimensionality"). There are promising directions for research at the interface of probabilistic/statistical modeling, computational modeling, high-performance computing, and application knowledge, suggesting that future research efforts in VVUQ should include collaborative interdisciplinary activities.

7.3.1 Verification Research

The solution verification process aims to quantitatively estimate the impact of numerical error on a given QOI. "Goal-oriented" methods are of particular interest, because they seek to estimate the error not in some abstract mathematical norm of the solution but rather in a given, defined functional of the solution—a particular QOI. As is discussed in Chapter 3, methods exist for estimating tight two-sided bounds for numerical error in the solution of linear elliptic partial differential equations (PDEs), but research is needed to develop a similar level of maturity for estimating error given more complicated mathematical models. In particular, the following areas of research have the potential for important practical improvements in verification methods.

- Development of goal-oriented a posteriori error-estimation methods that can be applied to mathematical models that are more complicated than linear elliptic PDEs. There are many such models that are of significant practical interest, including features such as nonlinearities, multiple coupled physical phenomena, bridging of multiple scales, hyperbolic PDEs, and stochasticity.
- Development of theory that supports goal-oriented error estimates on complicated grids, including adaptive mesh grids.
- Development of algorithms for goal-oriented error estimates that scale well on massively parallel architectures, especially given complicated grids (including adaptive mesh grids).

- Development of adaptive algorithms that can control numerical error given the kinds of complex mathematical models described above.
- Development of algorithms and strategies that efficiently manage both discretization error and iteration error, given the kinds of complex mathematical models described above.
- Development of methods to estimate error bounds when meshes cannot resolve important scales. An example is turbulent fluid flow.
- Further development of reference solutions, including "manufactured" solutions, for the kinds of complex mathematical models described above.
- For computational models that are composed of simpler components, including hierarchical models: development of methods that use numerical-error estimates from the simpler components, along with information about how the components are coupled, to produce numerical-error estimates for the overall model.

7.3.2 UQ Research

Although continued effort in improving methodology for building response surfaces and reduced-order models will likely prove fruitful in VVUQ, new research directions that consider VVUQ issues from a broader perspective are likely to yield more substantial gains in efficiency and accuracy. For example, response surface methods mentioned in Chapter 4 may consider both probabilistic descriptions of the input and the form of the mathematical/computational model to describe output uncertainty, leading to efficiency gains over standard approaches.

Embedded, or intrusive, approaches, such as those that use adjoint information for verification, sensitivity analyses, or inverse problems, tackle the problem from a perspective that leverages computational modeling aspects of the application, often achieving substantial gains in computational efficiency. In large-scale problems some approaches have folded in considerations regarding the computing architecture as well. However, beyond these examples, there is little in the current literature on how to exploit capabilities of high-performance computing in the service of VVUQ. The committee expects that VVUQ methodological research, operating from this broader perspective, will continue to be fruitful in the future.

Some applications use a collection of hierarchically connected models. In some cases, outputs from one model serve as inputs to another. Examples include the modeling of nuclear systems, or the reentry vehicle application described in Section 5.9. In other cases, a hierarchy of low- to high-fidelity computational models is available for modeling a particular system. An example is the modeling of radiative heat transfer using gray diffusion (low), multigroup diffusion (medium), or multi-group transport (high). In other cases, an application uses models that span multiple scales. In materials science, for example, where different models simulate phenomena at different scales ranging from molecular to mesoscale to large scale, where bulk properties such as strength emerge. In regional climate modeling, global and regional models are coupled to produce regional climate forecasts. In all of these cases there is opportunity to develop efficient approaches for VVUQ analyses that take advantage of a hierarchical structure.

There are challenges in such approaches, however. Liu et al. (2009) point out some of the obstacles that arise in the routine application of methodologies to link models. Determining how best to allocate resources for VVUQ investigations—an optimization problem—is an important UQ-related task that could benefit from further research. Optimization may take place rather narrowly, as in determining the best initial conditions over which to carry out a sequence of experiments, or more broadly, as in deciding between improving a module of the computational model or carrying out a costly experiment for a large VVUQ effort. Any such question requires some form of optimization while accounting for many sources of uncertainty.

The preceding paragraphs discuss areas in which improvements are needed in UQ methodology, and more detail is provided in Chapter 4. Here the committee summarizes some research directions that have the potential to lead to significantly improved UQ methods.

- Development of scalable methods for constructing emulators that reproduce the high-fidelity model results at training points, accurately capture the uncertainty away from training points, and effectively exploit salient features of the response surface.

- Development of phenomena-aware emulators, which would incorporate knowledge about the phenomena being modeled and thereby enable better accuracy away from training points (e.g., Morris, 1991).
- Exploration of model reduction for optimization under uncertainty.
- Development of methods for characterizing rare events, for example by identifying input configurations for which the model predicts significant rare events, and estimating their probabilities.
- Development of methods for propagating and aggregating uncertainties and sensitivities across hierarchies of models. (For example, how to aggregate sensitivity analyses across microscale, mesoscale, and macroscale models to give accurate sensitivities for the combined model remains an open problem.)
- Research and development in the compound area of (1) extracting derivatives and other features from large-scale computational models and (2) developing UQ methods that efficiently use this information.
- Development of techniques to address high-dimensional spaces of uncertain inputs. An important subset of problems is characterized by a large number of uncertain inputs that are correlated through subscale physical phenomena that are not included in the mathematical model being studied (an example of which is interaction coefficients in models involving particle transport).
- Development of algorithms and strategies, across the spectrum of UQ-related tasks, that can efficiently use modern and future massively parallel computer architectures.
- Development of optimization methods that can guide resource allocation in VVUQ while accounting for myriad sources of uncertainty.

7.3.3 Validation and Prediction Research

While many VVUQ tasks introduce questions that can be posed and answered (in principle) within the realm of mathematics, validation and prediction introduce questions whose answers require judgments from the realm of subject-matter expertise. It is challenging to quantify the effect of such judgments on VVUQ outcomes—that is, to translate them into the mathematical realm. This effort comes under the heading of assessing the quality of model-based predictions, which is a key research direction for improving the mathematical foundations of VVUQ.

For validation, "domain of applicability" is recognized as an important concept, but how one defines this domain remains an open question. For predictions, characterizing how a model differs from reality, particularly in extrapolative regimes, is a pressing need. While the literature has offered simple additive discrepancy models, as well as embedded, physically motivated discrepancy models (as in Box 5.1), advances in linking a model to reality will likely broaden the domain of applicability and improve confidence in extrapolative prediction.

Although multimodel ensembles offer an attractive pathway for assessing uncertainty due to model inadequacy, approaches to date have largely used ensembles of convenience, limiting their usefulness. While something is usually better than nothing, more rigorously constructed ensembles of models, designed so that reality is included within their spans, could ultimately provide a better foundation for assessing uncertainty. In a similar vein, some have advocated the use of more highly parameterized models to improve the chances of covering reality (Doherty and Welter, 2010), giving more realistic prediction uncertainties.

The use of large-scale computational models in searching out rare, high-consequence events, or estimating their probability, is particularly susceptible to discrepancies between model and reality. In such situations, models are almost always an extrapolation from available data, often extreme extrapolations. Here, too, research is needed.

The preceding paragraphs discuss areas in which improvements are needed in validation and prediction methodology, and more detail is provided in Chapter 5. Here the committee summarizes some research directions that have the potential to lead to significantly improved validation and prediction methods.

- Development of methods and strategies to quantify the effect of subject-matter judgments, which necessarily are involved in validation and prediction, on VVUQ outcomes.
- Development of methods that help to define the domain of applicability of a model, including methods that help to quantify the notions of near neighbors, interpolative predictions, and extrapolative predictions.
- Development of methods that incorporate mathematical, statistical, scientific, and engineering principles to produce estimates of uncertainty in "extrapolative" predictions.

- Development of methods or frameworks that help with the all-important problem of relating model-to-model differences, for models in an ensemble, to the discrepancy between models and reality.
- Development of methods to assess model discrepancy and other sources of uncertainty in the case of rare events, especially when validation data do not include such events.

Much of the research in this area should be a joint venture between subject-matter experts, mathematical/statistical experts, and computational modelers. The committee believes that the traditional funding plan in which funding is separated by field (mathematical sciences, computational science, basic science) is not ideal for making progress in the area of extrapolative predictions.

7.4 EDUCATION CHANGES FOR THE EFFECTIVE INTEGRATION OF VVUQ

The previous sections outline the current practices and future directions of VVUQ for large-scale computational simulations. Although scientists, engineers, and policy makers should of course use current best practices, there are important issues that have to be addressed to bring this about: (1) how to get the main concepts of VVUQ into the hands of those who need them so that the best practices become commonplace and (2) how to prepare the next generation of researchers. This section discusses educational changes in the mathematical sciences community that aim to integrate VVUQ and lay the foundation for improved methods and practices for the future.

As is discussed throughout this report, several broad tasks are included in VVUQ, and these tasks are likely to be performed by individuals with different areas of expertise. It is important that those involved understand these broad tasks and their implications. For instance, it is unlikely that a policy maker will carry out the task of code verification, but it is important that the person making the decisions understand the difference between a code that has gone through a VVUQ process and a code that has not. Conversely, it is equally important that computational modelers are cognizant of the potential uses of the computer code and that the predictive limitations of the computational model are clearly spelled out.

A report of the National Academy of Engineering (NAE), *The Engineer of 2020: Visions of Engineering in the New Century*, includes in its Executive Summary the vision of "improving our ability to predict risk and adapt systems" (NAE, 2004, p. 3). The same report describes the role of future engineers as continuing "to create solutions that minimize the risk of complete failure" (p. 24). In its report *Vision 2025*, the American Society of Civil Engineers (ASCE, 2006) describes civil engineers as (1) managers of risk and uncertainty caused by natural events, accidents, and other threats and (2) leaders in discussions and decisions shaping public environmental and infrastructure policy. It is reasonable to believe that similar characterizations enter the vision of other engineering and scientific disciplines.

A scientist or an engineer may have a part in several of the VVUQ tasks. Education plays an important role in making the best practices of VVUQ routine, and education and training should be targeted at the correct audiences, as is discussed further below.

7.4.1 VVUQ at the University

The development and implementation of VVUQ have been motivated by drivers similar to those underlying the NAE and ASCE visions. At the present time, topics in VVUQ are discussed at research conferences. Select topics come up in a few (usually graduate) engineering, statistics, and computer science courses, but a more encompassing view of VVUQ is not yet a standard part of the education of most undergraduate or graduate students. Because of the need to assess and manage risk and uncertainty within traditional mathematics-based modeling and to have confidence in the models for decision making, the educational objectives for VVUQ that could impact all undergraduate and graduate students in engineering, statistics, and the physical sciences should include the following:

- Probabilistic thinking,
- Science- and engineering-based modeling, and
- Numerical analysis and scientific computing.

Note that item 1 is included in some science and engineering programs, although it is often not required, and items 2 and 3 are not normally included in most probability and statistics or computer-science programs. With respect to items 1 and 2, it is necessary to identify mathematical tools relevant to applying probability and science together to address practical problems. With respect to items 2 and 3, it is necessary to understand how uncertainty can be introduced into deterministic physical laws and how evidence should be weighted to make model-based decisions.

It is reasonable to view VVUQ as motivating the need for an intellectual watershed merging items 1 through 3. VVUQ sits at the confluence of statistics, physics/engineering, and computing, which are themes that are usually discussed separately. To appreciate these distinctions, note that uncertainty is intimately associated with both observations and computational models. It is not about physical processes themselves but rather about one's interpretation (as embodied in mathematical models, assumptions, and uncertainty in data) of these processes. (If a physical process is random, it introduces another kind of uncertainty, which can be addressed within the mathematical model.) This perspective also holds for more empirically based models from fields such as operations research, psychology, and economics. Computational science is relevant to the extent that it permits the exploration of more detailed models, which helps to make better inferences about the real processes.

At the present time, undergraduate students are typically taught models of reality often without being introduced to the significance of the modeling process and without a critical assessment of associated assumptions and uncertainties. In engineering design courses, for example, students are most often introduced to a fait accompli in which a lack of knowledge and other uncertainties have already been integrated into a collection of safety factors. Students sometimes take advanced science and engineering courses before they have gained exposure to first principles of probability and statistics. Moreover, probability and statistics courses for engineering undergraduate students deal largely with data analysis (computing means, averages, point estimates, and confidence intervals) and do not introduce many concepts that are important to VVUQ.

A modern curriculum in UQ should equip its students with the foundation to reason about risks and uncertainty. This educational goal should include an understanding of the nature of risks associated with engineered and natural processes in an increasingly complex and interconnected world. Recent and ongoing events—including the nuclear reactor meltdown in Japan, the Deepwater Horizon blowout, engine failure in an Airbus 380 superjumbo jet, and the accelerated meltdown of ice sheets, among other examples—provide ready examples to motivate an understanding of the prevalence of risk. These problems are multifaceted and involve modeling themes from several traditional disciplines. A modern curriculum should foster an appreciation of the role that modeling and simulation could play in addressing such complex problems, providing clearer assessment of exposure, hazard, and risk and informing assessments of technical strategies for mitigating such hazards and risks. The curriculum should address effective communication of uncertainty and risk to decision makers, stakeholders, and UQ experts.

What might this mean for university programs? The required material to be integrated into an educational program will depend on the field. Students in engineering and science are routinely taught science- and engineering-based modeling and numerical methods and computing. Students of probability and statistics are taught probabilistic thinking and perhaps some numerical methods and computing. Decision makers (say, students of management) are likely to be introduced only to probabilistic thinking. The implications for different fields are briefly discussed below.

Recommendation: An effective VVUQ education should encourage students to confront and reflect on the ways that knowledge is acquired, used, and updated.

This recommendation can be achieved by assimilating relevant components of VVUQ as a fundamental scientific process into a minimal subset of core courses, sequenced in a manner that is conducive to the objective. Given the constraints of existing curricula, the alternative of integrating one or more new courses may not be feasible.

- *Engineering and science.* Any proposed educational program should respect the need for a logical sequence in knowledge acquisition. One can propose a route that first introduces the ubiquity of uncertainty throughout science and engineering. For example, this approach can be facilitated by the development of a number of examples that explain uncertainties associated with natural phenomena and engineering systems (e.g., the ball-drop examples in Chapters 1 and 5). This step can be followed by an introduction to probabilistic thinking, including classical as well as Bayesian statistics. Many of these ideas can likely be integrated into existing courses rather than requiring the introduction of new courses into an already-crowded curriculum. The engineering design process, as embodied in capstone design courses, can then be presented as a decision process aimed at selecting from competing alternatives, subject to various constraints. This formulation of the design process has the added benefit of articulating to the student the scientific distinctions between various design paradigms or procedures (usually presented in the form of design recipes). It may be that some programs already have such approaches, but they are not common.

- It is important to teach students to regularly confront uncertainty in input data and corresponding uncertainty in their stated answers. The committee encourages instructors in traditional courses to pose questions that include uncertainty in the input formation.
- Similar to engineers and scientists, students of probability and statistics should acquire training in mathematical modeling as well as in computational and numerical methods. Again, the path to doing so should build on the logical sequence of discipline-specific core training. The key is understanding how probabilistic thinking fits into the scientific process (e.g., how probability fits together with mathematical modeling) and also understanding the limits of computation.

Recommendation: The elements of probabilistic thinking, physical-systems modeling, and numerical methods and computing should become standard parts of the respective core curricula for scientists, engineers, and statisticians.

- *Programs in management sciences.* The intellectual framework represented by VVUQ seeks to assess the uncertainty in answers to a problem with respect to uncertainties in the given information. It is unlikely that students who are being trained as policy makers are going to be routinely interested in computational modeling, but it is important that they be educated in assessing the quality and reliability of the information that they are using to make decisions and also in assessing the inferential limits of the information. In the VVUQ context, this can mean, for example, understanding whether or not to trust a model that has undergone a VVUQ process, or understanding the distinction between predictions that have been informed by observations and those that have not.

It will be a challenge for individual university departments to take the lead in integrating VVUQ into their curricula. An efficient way of doing so would be to share the load among the relevant units. A way forward is emerging as a result of the DOE's Predictive Science Academic Alliance Program (PSAAP). For example, at the University of Michigan's Center for Radiative Shock Hydrodynamics (CRASH), both graduate and undergraduate students are included in the fundamental VVUQ steps as part of CRASH's core mission. More importantly in the current context, as a result of PSAAP, the university is initiating an interdisciplinary Ph.D. program in predictive science and engineering. Students in the program have a home department but will also take courses and develop methodology relating to VVUQ. (A course in VVUQ has already been taught.) The computational science, engineering, and mathematics program at the Institute for Computational and Engineering Sciences at the University of Texas also has a similar graduate program. It is not hard to imagine a similar interdisciplinary program (perhaps a certificate program in predictive science) being rolled out to undergraduate students in engineering, physics, probability and statistics, and possibly management science.

Finding: Interdisciplinary programs incorporating VVUQ methodology are emerging as a result of investment by granting bodies.

Recommendation: Support for interdisciplinary programs in predictive science, including VVUQ, should be made available for education and training to produce personnel who are highly qualified in VVUQ methods.

7.4.2 Spreading the Word

VVUQ plays an important role, with far-reaching consequences, in making sense of the information provided by computational models, observations, and expert judgment. It is important to communicate the best practices of VVUQ to those creating and using computational models and also to instructors in university programs.

To this end, several activities could be undertaken. For example, to provide assistance to instructors, some model problems and solutions have to be made available. In this spirit, people with expertise in the areas of VVUQ can be encouraged to write an article or series of articles targeted to an educational journal, in which problems are introduced and solutions are outlined.

Recommendation: Federal agencies should promote the dissemination of VVUQ materials and the offering of informative events for instructors and practitioners.

This type of contribution would go a long way toward sharing important ideas and suggesting how they might be implemented in a classroom setting. Along the same lines, the NAE could perhaps devote a special issue of its quarterly publication *The Bridge* to this type of initiative. It is also important to build on existing resources, such as the American Statistical Association's *Guidelines for Assessment and Instruction in Statistics Education*, which addresses the statistical component of VVUQ and highlights the need for a good understanding of data modeling, data analysis, data interpretation, and decisions. For existing practitioners, educational activities should be routinely included at conferences and also through the mathematical sciences institutes (e.g., the Statistical and Applied Mathematical Sciences Institute in Research Triangle Park, North Carolina, and the Mathematical Sciences Research Institute in Berkeley, California).

7.5 CLOSING REMARKS

This chapter attempts to peer into the future of VVUQ and to summarize the committee's responses to its tasking. It identifies key principles that we found to be helpful and identifies best practices that the committee has observed in the application of VVUQ to difficult problems in computational science and engineering. It identifies research areas that promise to improve the mathematical foundations that undergird VVUQ processes. Finally, it discusses changes in the education of professionals and dissemination of information that should enhance the ability of future VVUQ practitioners to improve and properly apply VVUQ methodologies to difficult problems, enhance the ability of VVUQ customers to understand VVUQ results and use them to make informed decisions, and enhance the ability of all VVUQ stakeholders to communicate with each other. These observations and recommendations are offered in the hope that they will help the VVUQ community as it continues to improve VVUQ processes and broaden their applications.

7.6 REFERENCES

ASCE (American Society of Civil Engineers). 2006. *Vision 2025*. Available at http://www.asce.org/uploadedFiles/Vision_2025. Accessed September 7, 2011.

Doherty, J., and D. Welter. 2010. A Short Exploration of Structural Noise. *Water Resources Research* 46:W05525.

Liu, F., M.J. Bayarri, and J. Berger. 2009. Modularization in Bayesian Analysis, with Emphasis on Analysis of Computer Models. *Bayesian Analysis* 4:119-150.

Morris, M. 1991. Factorial Sampling Plans for Preliminary Computational Experiments. *Technometrics* 33(2):161-174.

NAE (National Academy of Engineering). 2004. *The Engineer of 2020: Visions of Engineering in the New Century*. Washington, D.C.: The National Academies Press.

Oreskes, N., K. Shrader-Frechette, and K. Berlitz. 1994. Verification, Validation, and Confirmation of Numerical Models in the Earth Sciences. *Science* 263(5147):641-646.

Randall, D.A., R.A. Wood, S. Bony, R. Colman, T. Fichefet, J. Fyfe, V. Kattsov, A. Pitman, J. Shukla, J. Srinivasan, R.J. Stouffer, A. Sumi, and K.E. Taylor. 2007. Climate Models and Their Evaluation. Pp. 591-648 in *Climate Change 2007: The Physical Science Basis. Contribution of Working Group I to the Fourth Assessment Report of the Intergovernmental Panel on Climate Change*. S.D. Solomon, D. Qin, M. Manning, Z. Chen, M.C. Marquis, K.B. Averyt, M. Tignor, and H.L. Miller (Eds.). Cambridge, U.K.: Cambridge University Press.

Appendixes

Appendix A

Glossary

TABLE A.1 Glossary of Terms Related to Verification, Validation, and Uncertainty Quantification

Term, with Synonyms and Cross-References	Definition	Notes and Comments
accuracy See also **precision**.	A measure of agreement between the estimated value of some quantity and its true value. (Adapted from Society for Risk Analysis [SRA] Glossary.[a])	See note under **precision**.
adjoint map	Given a map (i.e., forward model) from an input vector space to an output vector space, the adjoint is an associated map between the vector space of linear real-valued functions on the output space to the vector space of linear real-valued functions on the input space. Given a linear real function on the output space, a linear real function on the input space is obtained by first applying the original map to any specified vector in the input space and then applying the given linear real function on the output space.	The adjoint map is important for determining properties of the original map when the input and output vectors cannot be observed directly. It plays a fundamental role in the theory of maps, e.g., for determining solvability of inverse problems, stability and sensitivity, Green's functions, and derivatives of the output of a map with respect to the input. The concrete formulation and evaluation of an adjoint depend heavily on the properties of the original map (i.e., forward model) and the input and output spaces, with extra care needed for nonlinear maps.
aleatoric uncertainty Synonyms: aleatoric probability, aleatoric uncertainty, systematic error See also **probability, epistemic uncertainty**.	A measure of the **uncertainty** of an unknown event whose occurrence is governed by some random *physical* phenomena that are either (1) predictable, in principle, with sufficient information (e.g., tossing a die), or (2) essentially unpredictable (radioactive decay).[b]	See **epistemic uncertainty**.

109

Term, with Synonyms and Cross-References	Definition	Notes and Comments
algorithm	A finite list of well-defined instructions that, when executed, proceed through a finite number of well-defined successive states, eventually terminating and producing an output.	The instructions and executions are not necessarily deterministic; some algorithms incorporate random input (see **Monte Carlo simulation**).
approximation See also **estimation (of parameters in probability models)**.	The result of a computation or assessment that may not be exactly correct but that is adequate for a particular purpose.[c]	
average Synonyms: arithmetic mean, sample mean See also **mean**.	The sum of n numbers divided by n.[d,e,f]	The average is a simple arithmetic operation requiring a set of n numbers. It is often confused with the **mean** (or **expected value**), which is a property of a **probability distribution**. One reason for this confusion is that the average of a set of realizations of a random variable is often a good estimator of the mean of the random variable's distribution.
Bayesian approach See also **prior probability**.	An approach that uses observations (data) to constrain uncertain parameters in a probabilistic model. The constrained uncertainty is described by a posterior probability distribution, produced using Bayes's theorem to combine the prior probability distribution with the probabilistic model of the observations.	In most problems the Bayesian approach produces a high-dimensional **probability distribution** describing the joint uncertainty in all of the model parameters. Functionals or integrals of this posterior distribution are typically used to summarize the posterior uncertainty. These summaries are typically produced by means of numerical approximation or sampling methods such as **Markov chain Monte Carlo**.
code verification See also **verification, solution verification**.	The process of determining and documenting the extent to which a computer program ("code") correctly solves the equations of the mathematical model.	
computational model Synonym: computer model See also **model (simulation)**.	Computer code that (approximately) solves the equations of the mathematical model.	In physically based applications the computational model might encode physical rules such as conservation of mass or momentum. In other applications the computational model might also produce a Monte Carlo or a discrete-event realization.
conditional probability See also **probability**.	The **probability** of an event supposing (i.e., "conditioned on") the occurrence of other specified events.	In the Bayesian approach the posterior distribution is a conditional probability distribution, conditioned on the physical observations. It is important to note that subjectively assessed probabilities are based on the state of knowledge that holds at the time of the probability assessment.

APPENDIX A

Term, with Synonyms and Cross-References	Definition	Notes and Comments
confidence interval Synonym: interval	A range of values $[a, b]$ determined from a sample, using a predetermined rule chosen such that, in repeated random samples from the same population, the fraction α of computed ranges will include the true value of an unknown parameter. The values a and b are called confidence limits; α is called the confidence coefficient (commonly chosen to be .95 or .99); and $1 - \alpha$ is called the confidence level. (Adapted from SRA Glossary.)[a]	Confidence intervals should not be interpreted as implying that the parameter itself has a range of values; it has only one value. For any given sample the confidence limits a and b define a random range within which the parameter of interest will lie with probability a (provided that the actual population satisfies the initial hypothesis).
constrained uncertainty See also **Bayesian approach**.	**Uncertainty** about a parameter, prediction, or other entity that has been reduced by incorporating additional information, such as new physical observations.	For most of the examples in this report, uncertainty is constrained using the Bayesian approach, conditioning on physical observations, producing a posterior distribution for parameters and predictions.
continuous random variable See also **cumulative distribution function**, **probability density function**.	A random variable, X, is continuous if it has an absolutely continuous **cumulative distribution function**.[d]	
cumulative distribution function Synonyms: cumulative distribution, cdf, distribution function See also **probability density function**, **probability distribution**.	The **probability** that a random variable X will be less than or equal to a value x; written as $P\{X \leq x\}$.[f,g]	The cdf always exists for any random variable; it is monotonic nondecreasing in x, and (being a probability $0 \leq P\{X \leq x\} \leq 1$. If $P\{X \leq x\}$ is absolutely continuous in x, then X is called a **continuous random variable**; if it is discontinuous at a finite or countably infinite number of values of x, and constant otherwise, X is called a **discrete random variable**.
data assimilation	A recursive process for producing predictions with **uncertainty** regarding some process, commonly used in weather forecasting and other fields of geoscience. At a given iteration, new physical observations are combined with model-based predictions to produce updated predictions and updated estimates of the current state of the system.	The combination method is usually based on Bayesian inference. The Kalman filter, the ensemble Kalman filter, and particle filters are examples of approaches with which data assimilation is carried out.
data verification and validation	The process of verifying the internal consistency and correctness of data and validating that they represent real-world entities appropriate for their intended purpose or an expected range of purposes.[h]	
discrete random variable See also **cumulative distribution function**.	A random variable that has a nonzero probability for only a finite, or countably infinite, set of values.[b]	

Term, with Synonyms and Cross-References	Definition	Notes and Comments
epistemic uncertainty Synonym: epistemic probability See also **aleatoric uncertainty**.	A representation of **uncertainty** about propositions due to incomplete knowledge. Such propositions may be about either past or future events.[b]	Some examples of epistemic uncertainty are (1) a **probability density function** describing uncertainty regarding the acceleration due to gravity at Earth's surface; (2) determination of the probability that a required maintenance procedure will, in fact, be carried out.
estimation (of parameters in probability models) See also **approximation**.	A procedure by which sample data are used to assess the value of an unknown quantity.[e]	Estimation procedures are usually based on statistical analyses that address their efficiency, effectiveness, limiting behaviors, degrees of bias, etc. The most common methods of parameter estimation are "maximum likelihood" and the method of moments. Under the **Bayesian approach** estimates can be produced by taking the **mean**, median, or most likely value determined by the posterior distribution.
expected value Synonym: expectation See also **mean**.	The first moment of the probability distribution of a random variable X; often denoted as $E(X)$ and defined as $\sum x_i p(x_i)$ if X is a **discrete random variable** and $\int x f(x) dx$ if X is a **continuous random variable**.[d,f]	
extrapolative prediction See also **interpolative prediction**.	The use of a model to make statements about **quantities of interest** (QOIs) in settings (initial conditions, physical regimes, parameter values, etc.) that are outside the conditions for which the model validation effort occurred.	
face validation See also **validation**.	A nonquantitative "sanity check" on a model that requires both its structural content and outputs to be consistent with well-understood and agreed-on forms, ranges, etc.	Face validation should not be used by itself as a formal validation process. Instead, it should be used to guide model development, design of sensitivity analyses, etc.
forward problem See also **inverse problem**.	The use of a model, given the values of all necessary inputs (initial conditions, parameters, etc.), to produce potentially observable **QOIs**.	
forward propagation Synonym: uncertainty propagation (UP) See also **forward problem**.	Quantifying the **uncertainty** of a model's responses that results from uncertainty in the model's inputs being propagated through the model.	

APPENDIX A

Term, with Synonyms and Cross-References	Definition	Notes and Comments
global statistical sensitivity analysis See also **sensitivity analysis**.	The study of how the **uncertainty** in the output or **QOI** of a model (numerical or otherwise) can be apportioned to different sources of uncertainty in the model input. The term *global* ensures that the analysis considers more than just local or one-factor-at-a-time effects. Hence interactions and nonlinearities are important components of a global statistical sensitivity analysis.	Global statistical sensitivity analysis is distinguished from local, or one-at-a-time, sensitivity analyses in that interactions and nonlinearities are considered.
input verification See also **verification**.	The process of determining that the data entered into a model or **simulation** accurately represent what the developer intends. (Adapted from DOD, 2009.[h])	
interpolative prediction See also **extrapolative prediction**.	The use of a model to make statements about **QOIs** in regimes within which the model has been validated.	In practice, it may be difficult to determine if a particular prediction is an interpolation or not.
intrusive methods See also **nonintrusive methods (black box methods)**.	Approaches to exploring a **computational model** that require a recoding of the model. Such a recoding might be done in order to efficiently produce derivative information using the **adjoint equation** to facilitate a **sensitivity analysis**.	
inverse problem See also **forward problem**.	An estimation of a model's uncertain parameters by using data, measurements, or observations.	An inverse problem is often formulated as an optimization problem that minimizes an appropriate measure of the "differences" between observed and model-predicted outputs (with constraints—or penalty costs—on the values of some of the parameters).
level of fidelity See also **validation**.	The amount of detail with which a model describes an actual process. Relevant features might include the descriptions of geometry, model symmetries, dimensionality, or physical processes in the model. High-fidelity models attempt to capture more of these features than do low-fidelity models.	A high level of fidelity does not necessarily imply that the model will give highly accurate predictions for the system.
likelihood See also **probability, uncertainty**.	The likelihood, $L(A \mid D)$, of an event, A, given the data, D, and a specific model, is often taken to be proportional to $P(D \mid A)$, the constant of proportionality being arbitrary.[i]	In informal usage, "likelihood" is often a qualitative description of **probability** or frequency. However, equally often these descriptions do not satisfy the axioms of probability.
linear regression Synonym: regression See also **nonlinear regression**.	Regression when the function to be fit is linear in the independent variables.	

Term, with Synonyms and Cross-References	Definition	Notes and Comments
Markov chain Monte Carlo (MCMC)	A sampling technique that constructs a Markov chain to produce Monte Carlo samples from a typically complicated, multivariate distribution. The resulting sample is then used to estimate functionals of the distribution.	MCMC typically requires many fewer points than grid-based sampling methods. MCMC approaches become intractable as the complexity of the forward problem and the dimensions of the parameter spaces increase.
mathematical model Synonym: conceptual model See also **model (simulation)**.	A model that uses mathematical language (sets of equations, inequalities, etc.) to describe the behavior of a system.	
mean See also **expected value, average**.	The first moment of a **probability distribution**, with the same mathematical definition as that of **expected value**. The mean is a parameter that represents the central tendency of a distribution.[d,e,g,j]	
measurement error	The discrepancy between a measurement and the quantity that the measurement instrument is intended to measure.[k]	Measurement error is often decomposed into two components: replicate variation and bias.
model (simulation) See also **simulation**.	A representation of some portion of the world in a readily manipulated form. A **mathematical model** is an abstraction that uses mathematical language to describe the behavior of a system.[l]	Mathematical models are used to aid our understanding of some aspects of the real world and to aid in decision making. They are also valuable rhetorical tools for presenting the rationale supporting various decisions, since they arguably allow for transparency and the reproduction of results by others. However, models are only as good as their (validated) relationship to the real world and within the context for which they are designed.
model discrepancy Synonyms: model inadequacy, structural error	A term accounting for or describing the difference between a model of the system and the true physical system.	In some cases, model discrepancy is the dominant source of **uncertainty** in model-based predictions. When relevant physical data are available, model discrepancy can be estimated. Estimating this term when relevant physical observations are not available is difficult.
Monte Carlo simulation See also **model (simulation)**.	A model constructed so that the input of a large number of random draws from defined **probability distributions** will generate outputs that are representative of the random behavior of a particular system, phenomenon, consequences, etc., of a series of events.[m]	Each set of "runs" of a **simulation** inherently represents the outcomes of a series of experiments. The analysis of simulation output data therefore requires a proper experimental design, followed by the use of statistical techniques to estimate parameters, test hypotheses, etc.
multiscale phenomena	Equations representing the dynamics of a nonlinear system that combine the behavior at many scales of physical dimension and/or time.	The analysis of multiscale phenomena presents many challenges to numerical analysis and associated software, so that the coupling of results from one scale to those of another may lead to instability in the model output that might not represent physical reality.

Term, with Synonyms and Cross-References	Definition	Notes and Comments
multivariate adaptive regression splines (MARS) See also **regression**.	A form of nonparametric regression analysis (usually presented as an extension of **linear regression**) that automatically represents nonlinearities and interactions in terms of splines (e.g., functions having smooth first and second derivatives).[n]	
nonintrusive methods (black box methods)	Methods to carry out **sensitivity analysis** or **forward propagation** or to solve the **inverse problem** that only require forward runs of the **computational model**, effectively treating the model as a black box.	
nonlinear regression See also **regression**, **linear regression**.	Regression when the function to be fit is nonlinear in the independent variables.	
parameter	Terms in a mathematical function that remain fixed during any computational procedure. These may include initial conditions, physical constants, boundary values, etc.	Often parameters are fixed at assumed values, or they can be estimated using physical observations. Alternatively, uncertainty regarding parameters may be constrained with physical data.
polynomial chaos Synonym: PC, Wiener chaos expansion See also **Monte Carlo simulation**.	A parameterization of random variables and processes that lends itself to the characterization of transformations between input and output quantities. The resulting representations are akin to a **response surface** with respect to normalized random variables and can be readily evaluated, yielding very efficient procedures for sampling the output variables.	The coefficients in these representations can be estimated in a number of ways, including Galerkin projections, least squares, perturbation expansions, statistical sampling, and numerical quadrature.
posterior probability See also **Bayesian approach, prior probability**.	**Probability distribution** describing uncertainty in parameters (and possibly other random quantities) of interest in a statistical model *after* data are observed and conditioned on.	The **Bayesian approach** updates the **prior probability** distribution by conditioning on the data (often physical observations), producing a posterior distribution for the same parameters. Often of interest is the posterior predictive distribution for a **QOI**, describing **uncertainty** about the QOI for the physical system.
precision See also **accuracy**.	The implied degree of certainty with which a value is stated, as reflected in the number of significant digits used to express the value—the more digits, the more precision. (Adapted from SRA Glossary.[a])	Consider two statements assessing Bill Gates's net worth, W. A precise but inaccurate assessment is "$W = \$123,472.89$." An imprecise but accurate assessment is "$W > \$6$ billion."

Term, with Synonyms and Cross-References	Definition	Notes and Comments
prediction uncertainty	The **uncertainty** associated with a prediction about a **QOI** for the real-world process. The prediction uncertainty could be described by a posterior distribution for the QOI, a predictive distribution, a confidence interval, or possibly some other representation.	This is a statement about reality, given information from an analysis typically involving a **computational model**, physical observations, and possibly other information sources.
prior probability Synonym: a priori probability See also **Bayesian approach, posterior probability.**	Probability distribution assigned to parameters (and possibly other random quantities) of interest in a statistical model *before* physical observations are available.	**Bayesian approach** updates this prior probability distribution by conditioning on the physical observations, producing a posterior distribution for the same parameters. Obtaining the prior distribution may be done using expert judgment or previous data, or it may be specified to be "neutral" to the analysis.
probability See also **likelihood, conditional probability, aleatoric uncertainty, subjective probability.**	One of a set of numerical values between 0 and 1 assigned to a collection of random events (which are subsets of a sample space) in such a way that the assigned numbers obey two axioms: (1) $0 \leq P\{A\} \leq 1$ for any A and (2) $P\{A\} + P\{B\} = P\{A \cup B\}$ for two mutually exclusive events A and B.[j]	This definition holds for *all* quantification of **uncertainty**: subjective or frequentist.
probability density function (pdf)	The derivative of an absolutely continuous cumulative distribution function.[j] For a scalar random variable X, a function f such that, for any two numbers, a and b, with $a \leq b$, $P\{a \leq X \leq b\} = \int_a^b f(x)dx$.	The **pdf** is the common way to represent the **probability distribution** of a **continuous random variable**, because its shape often displays the central tendency (mean) and variability (**standard deviation**). From its definition, $P\{a < X \leq b\}$ is the integral of the **pdf** between a and b.
probability distribution	See **cumulative distribution function**.	
probability elicitation Synonyms: probability assessment, subjective probability	A process of gathering, structuring, and encoding expert judgment (about uncertain events or quantities) in the form of probability statements about future events.[o]	There are many approaches for probability elicitation, the most common of which are those used for obtaining a priori subjective probabilities. Note that the results of probability elicitations are sometimes called probability assessments or assignments.
quantity of interest (QOI)	A numerical characteristic of the system being modeled, the value of which is of interest to stakeholders, typically because it informs a decision. To be useful the model must be able to provide, as output, values of or probability statements about QOIs.	

APPENDIX A

Term, with Synonyms and Cross-References	Definition	Notes and Comments
reduced model Synonym: emulator	A low-fidelity model developed to replace (or augment) a computationally demanding, high-fidelity model.	A reduced model is particularly useful for carrying out computationally demanding analysis (e.g., **sensitivity analysis**, **forward propagation** of uncertainty, solving the **inverse problem**) that would be infeasible with the original model. Sometimes a reduced model "collapses" aspects of a "physics-based" model so as to be referred to as a "physics-blind" model.
regression See also: **linear regression**, **nonlinear regression**.	A form of statistical analysis in which observational data are used to statistically fit a mathematical function that presents the data (i.e., dependent variables) as a function of a set of parameters and one or more independent variables.	
response surface See also **sensitivity analysis**.	A function that predicts outputs from a model as a function of the model inputs. A response surface is typically estimated from an ensemble of model runs using a regression, Gaussian process modeling, or some other estimation or interpolation procedure.	A response surface can be used like a **reduced model** to carry out computationally demanding analyses (e.g., **sensitivity analysis**, **forward propagation**, solving the **inverse problem**). Since the response surface does not exactly reproduce the **computational model**, there is typically additional error in results produced by response surface approaches.
robustness analysis See also **sensitivity analysis**.	For a prescriptive model, a procedure that analyzes the degree to which deviations from a "best" decision provide suboptimal values of the desired criterion. These deviations can be due to uncertainty in model formulation, assumed parameter values, etc.	
sensitivity analysis See also **robustness analysis**.	An exploration, often by numerical (rather than analytical) means, of how model outputs (particularly **QOIs**) are affected by changes in the inputs (parameter values, assumptions, etc.).	
simulation Synonym: model See also **Monte Carlo simulation**.	The execution of a computer code to mimic an actual system.	Many **uncertainty quantification** (UQ) methods use an ensemble of simulations, or model runs, to construct emulators, carry out **sensitivity analysis**, etc.
solution verification See also **verification**, **code verification**.	The process of determining as completely as possible the accuracy with which the algorithms solve the mathematical-model equations for a specified **QOI**.	
standard deviation See also **variance**.	The square root of the **variance** of a distribution.[j]	

Term, with Synonyms and Cross-References	Definition	Notes and Comments
stochastic See also **probability**.	Pertaining to a sequence of observations, each of which can be considered to be a sample from a **probability distribution**.	Often informally used as a synonym of "probabilistic."
subjective probability See also **probability elicitation**.	Expert judgment about uncertain events or quantities, in the form of probability statements about future events. It is not based on any precise computation but is often a reasonable assessment by a knowledgeable person.	
uncertainty See also **probability, aleatoric probability, epistemic uncertainty.**	The condition of being unsure about something; a lack of assurance or conviction.[c]	For the purpose of this report, uncertainty is often described regarding a **QOI** of the true, physical system. This uncertainty depends on a model-based prediction, as well as on other information included in the VVUQ assessment. This uncertainty can be described using **probability**.
uncertainty quantification (UQ)	The process of quantifying uncertainties in a computed **QOI**, with the goals of accounting for all sources of **uncertainty** and quantifying the contributions of specific sources to the overall uncertainty.	More broadly, UQ can be thought of as the field of research that uses and develops theory, methodology, and approaches for carrying out inference, with the aid of **computational models**, on complex systems.
validation	The process of determining the degree to which a model is an accurate representation of the real world from the perspective of the intended uses of the model.[D]	
variance See also **standard deviation**.	The second moment of a **probability distribution**, defined as $E(X - \mu)^2$, where μ is the first moment of the random variable X.	The **variance** is a common measure of variability around the mean of a distribution. Its square root, the **standard deviation**, having dimensional units of the random variable, is a more intuitively meaningful measure of dispersion from the mean.
verification See also **code verification, solution verification**.	The process of determining whether a computer program ("code") correctly solves the mathematical-model equations. This includes **code verification** (determining whether the code correctly implements the intended algorithms) and **solution verification** (determining the accuracy with which the algorithms solve the mathematical-model equations for specified QOIs).	

[a] Society for Risk Analysis (SRA), Glossary of Risk Analysis Terms. Available at sra.org/resources_glossary.php.
[b] Cornell LCS Statistics Laboratory. See http://instruct1.cit.cornell.edu:8000/courses/statslab/Stuff/indes.php.
[c] *American Heritage Dictionary*. 2000. Boston: Houghton, Mifflin.
[d] Glossary of Statistics Terms. Available at http://www.stat.berkeley.edu/ users/stark/SticiGui/Text/gloss.htm.
[e] Statistical Education Through Problem Solving [STEP] Consortium. Available at http://www.stats.gla.ac.uk/steps/index.html.
[f] W. Feller. 1968. *An Introduction to Probability Theory and Its Applications*. New York, N.Y.: Wiley.
[g] J.L. Devore. 2000. *Probability and Statistics for Engineering and the Sciences*. Pacific Grove, Calif.: Duxbury Press.

[h] DOD (Department of Defense). 2009. Instruction 5000.61. December 9. Washington, D.C.
[i] A.W.F. Edwards. 1992. *Likelihood*. Baltimore, Md.: Johns Hopkins University Press.
[j] S.M. Ross. 2000. *Introduction to Probability Models*. New York: Academic Press.
[k] Duke University. 1998. *Statistical and Data Analysis for Biological Sciences*. Available at http://www.isds.duke.edu/courses/Fall98/sta210b/terms.html.
[l] R. Aris. 1995. *Mathematical Modelling Techniques*, New York: Dover.
[m] E.J. Henley and H. Kunmamoto. 1981. *Reliability Engineering and Risk Assessment*. Upper Saddle River, N.J.: Prentice-Hall.
[n] J.H. Friedman. 1991. Multivariate Adaptive Regression Splines. *The Annals of Statistics* 19(1):1-67.
[o] M.S. Meyer and J.M. Booker. 1998. *Eliciting and Analyzing Expert Judgment*. LA-UR-99-1659. Los Alamos, N.Mex.: Los Alamos National Laboratory.
[p] American Institute for Aeronautics and Astronautics. 1998. *Guide for the Verification and Validation of Computational Fluid Dynamics Simulations*. Reston, Va.: American Institute for Aeronautics and Astronautics.

Appendix B

Agendas of Committee Meetings

JUNE 10-11, 2010
ALBUQUERQUE, NEW MEXICO

June 10, 2010

CLOSED SESSION

8:00 a.m.	Working Breakfast	
8:30	Introduction Bias and Conflicts of Interest Discussion	Neal Glassman, National Research Council (NRC)

OPEN SESSION

9:30 a.m.	Verification and Validation/Uncertainty Quantification (V&V/UQ) for the Department of Energy (DOE) National Security Mission	Thuc Hoang, DOE/National Nuclear Security Administration (NNSA)
10:15	Break	
10:30	A Brief View of Verification, Validation, and Uncertainty Quantification	J. Tinsley Oden, University of Texas at Austin
11:15	Review of NRC Report on Quantification of Margins of Uncertainty	Marvin L. Adams, *Committee Co-Chair*, Texas A&M University
Noon	Working Lunch	

APPENDIX B

1:00 p.m.	Opinions on V&V	Timothy Trucano, Sandia National Laboratories
1:45	The Advance of UQ Science: Challenges and Approaches	Richard Klein, Lawrence Livermore National Laboratory
2:35	Break	
2:50	What Is Predictive Capability?	Mark Anderson, Los Alamos National Laboratory
3:35-4:20	Overview of Uncertainty Quantification: Research and Deployment in the DAKOTA project	Michael Eldred, Sandia National Laboratories

CLOSED SESSION

4:15 p.m.	Committee Discussion	
6:00	Working Dinner	

June 11, 2010

8:00 a.m.	Working Breakfast	

OPEN SESSION

8:30 a.m.	The Virtual Measurement Systems Program at the National Institute of Standards and Technology (NIST)	Andrew Dienstfrey, NIST
9:00	Uncertainty Quantification and V&V for Environmental Models	Bruce Robinson, Los Alamos National Laboratory
9:45	Break	
10:00	Climate Model Uncertainties	Peter Gleckler, Lawrence Livermore National Laboratory

CLOSED SESSION

10:45 a.m.	Committee Discussion	

AUGUST 23-24, 2010
WASHINGTON, D.C.

August 23, 2010

CLOSED SESSION

8:00 a.m.	Working Breakfast	
8:30	Reprise of Bias and Conflicts of Interest Discussion	Neal Glassman, NRC

OPEN SESSION — David M. Higdon, *Committee Co-Chair*, Los Alamos National Laboratory

8:45 a.m.	Uncertainty, Risk, and Expert Opinion	Bilal Ayyub, University of Maryland
9:45	Verification and Validation Methods	Mikel Petty, University of Alabama in Huntsville
10:45	Break	
11:00	V&V in Warfare Modeling	Susan Sanchez, Naval Postgraduate School
Noon	Working Lunch	

OPEN SESSION — Marvin L. Adams, *Committee Co-Chair*, Texas A&M University

12:45 p.m.	V&V in Very Large-Scale Simulation	Christopher Barrett, Virginia Polytechnic Institute and State University
1:45	V&V in Economic Modeling	Christopher Sims, Princeton University
2:45	Break	
3:00	V&V in the Multi-Modeling Domain	Alex Levis, George Mason University

CLOSED SESSION

4:00 p.m.	Discussion of Issues and Report
5:00	Reception
6:00	Working Dinner

APPENDIX B

August 24, 2010

OPEN SESSION David M. Higdon, *Committee Co-Chair*,
Los Alamos National Laboratory

8:00 a.m. Working Breakfast

8:30 Limits of Predictability Louis Anthony Cox, Jr., Cox Associates

9:30 Modeling at the University of Illinois at Eric Michielssen, University of Michigan
Urbana-Champaign

CLOSED SESSION

10:30 a.m. Working Lunch and Further Discussion

FEBRUARY 1-2, 2011
IRVINE, CALIFORNIA

The meeting was closed in its entirety.

MARCH 17-18, 2011
SAN FRANCISCO, CALIFORNIA

The meeting was closed in its entirety.

MAY 18-19, 2011
CHICAGO, ILLINOIS

The meeting was closed in its entirety.

JULY 8-9, 2011
LA JOLLA, CALIFORNIA

The meeting was closed in its entirety.

Appendix C

Committee Biographies

MARVIN L. ADAMS, *Co-Chair,* is the HTRI [Heat Transfer Research, Inc.] Professor of Nuclear Engineering and director of the Institute for National Security Education and Research at Texas A&M University. His research has focused on many aspects of computational science and engineering, including discretization methods, iterative methods, parallel algorithms, and the quantification of predictive capability. He has served as a consultant to the Lawrence Livermore National Laboratory (LLNL), Sandia National Laboratories, and Los Alamos National Laboratory and has served on a variety of review and advisory committees and panels for the laboratories, the Department of Energy, and other governmental organizations. Dr. Adams earned his B.S. (1981) from Mississippi State University followed by M.S. (1984) and Ph.D. (1986) degrees from the University of Michigan, all in nuclear engineering. From 1977 to 1982 he worked at the Tennessee Valley Authority's Sequoyah Nuclear Plant and its support office. He joined LLNL after completing his graduate work in 1986. He left LLNL in 1992 for the faculty position that he continues to hold at Texas A&M University. In 2006 and 2007, Dr. Adams founded and directed the Center for Large-scale Scientific Simulations at Texas A&M, and from 2005 until 2009 he served as associate vice president for research. Previously Dr. Adams served on the National Research Council's Committee on Evaluation of Quantification of Margins and Uncertainties Methodology for Assessing and Certifying the Reliability of the Nuclear Stockpile.

DAVID M. HIGDON, *Co-Chair,* is a member of the Statistical Sciences Group at the Los Alamos National Laboratory (LANL). He is an internationally recognized expert in Bayesian statistical modeling of environmental and physical systems. He has also led numerous programmatic efforts at LANL in the quantification of margins and uncertainties and uncertainty quantification. His recent research has focused on simulation-aided inference in which physical observations are combined with computer simulation models for prediction and inference. His research interests include space-time modeling; inverse problems in physics, hydrology, and tomography; inference based on combining deterministic and stochastic models; multiscale models; parallel processing in posterior exploration; statistical modeling in physical, environmental, and biological sciences; and Monte-Carlo and simulation-based methods.

JAMES O. BERGER is Arts and Sciences Professor of Statistics at Duke University. He was a faculty member in the Department of Statistics at Purdue University until 1997, when he moved to the Institute of Statistics and Decision Sciences (now the Department of Statistical Science) at Duke University. He has also been the director of the national Statistical and Applied Mathematical Sciences Institute since 2002. He was the president of the Institute of Mathematical Statistics (1995-1996), chair of the Section on Bayesian Statistical Science of the American Statistical Association (1995), and president of the International Society for Bayesian Analysis (2004). He has been involved in numerous editorial activities, including co-editorship of the *Annals of Statistics* (1998-2000). He has organized or participated in the organization of more than 35 conferences. Among the awards and honors that Professor Berger has received are Guggenheim and Sloan Fellowships, the COPSS [Committee of Presidents of Statistical Societies] President's Award (1985), the Sigma Xi Research Award at Purdue University for contribution of the year to science (1993), the Fisher Lectureship (2001), election as a foreign member of the Spanish Real Academia de Ciencias (2002), election to the U.S. National Academy of Sciences (2003), an honorary doctor of science degree from Purdue University (2004), and the Wald Lectureship (2007). Professor Berger's research has been primarily in Bayesian statistics, foundations of statistics, statistical decision theory, simulation, model selection, and various interdisciplinary areas of science and industry, especially astronomy and the interface between computer modeling and statistics. He has supervised 31 Ph.D. dissertations, published more than 160 articles, and written or edited 14 books or special volumes.

DEREK BINGHAM is an associate professor and the Canada Research Chair in Industrial Statistics in the Department of Statistics and Actuarial Science at Simon Fraser University. He received his Ph.D. from the Department of Mathematics and Statistics at Simon Fraser University in 1999. After graduation he joined the Department of Statistics at the University of Michigan as an assistant professor, returning to Simon Fraser in 2003. In addition, he has held a faculty affiliate position at the Los Alamos National Laboratory. The main focus of Dr. Bingham's research is the development of statistical methodology for the design and analysis of industrial and physics experiments. This work focuses on developing new methodology for (1) the design and analysis of computer experiments and (2) the design and analysis of experiments in industrial problems such as optimal screening designs, response surface optimization, and optimal robust parameter designs for product variation reduction.

WEI CHEN is the Wilson-Cook Chair and Professor in Engineering Design at Northwestern University. She is affiliated with the Segal Design Institute as a faculty fellow and is a professor in the Department of Mechanical Engineering, with a courtesy appointment in the Department of Industrial Engineering and Management. As a director of the Integrated Design Automation Laboratory, her current research involves issues such as simulation-based design under uncertainty, model validation, stochastic multiscale analysis and design, robust shape and topology optimization, multidisciplinary optimization, consumer choice modeling, and enterprise-driven decision-based design. She is the co-founder and director of the interdisciplinary doctoral cluster in predictive science and engineering design at Northwestern University, a program aiming for integrating scientific, physics-based modeling, and simulation into the design of innovative "engineered" systems. Dr. Chen is the recipient of a 1996 National Science Foundation Faculty Early Career Award and the 1998 American Society of Mechanical Engineers (ASME) Pi Tau Sigma Gold Medal achievement award. She is also a recipient of the 2005 Intelligent Optimal Design Prize and the 2006 Society of Automotive Engineering (SAE) Ralph R. Teetor Educational Award. Dr. Chen is a fellow of ASME, an associate fellow of the American Institute of Aeronautics and Astronautics, and a member of SAE. She is an elected member of the ASME Design Engineering Division Executive Committee and an elected Advisory Board member of the Design Society, an international design research community. She is an associate editor of the ASME *Journal of Mechanical Design* and serves as the review editor of *Structural and Multidisciplinary Optimization*. In the past, she served as the chair and a member of the ASME Design Automation Executive Committee (2002-2007) and was an associate editor of the *Journal of Engineering Optimization*.

ROGER GHANEM is the Gordon S. Marshall Professor of Engineering Technology in the Viterbi School at the University of Southern California (USC). Dr. Ghanem has a Ph.D. in civil engineering from Rice University and had served on the faculty of the Schools of Engineering at SUNY [State University of New York]-Buffalo and Johns Hopkins University before joining USC in 2005. Dr. Ghanem's research is mainly in the area of computational science and engineering with a focus on uncertainty quantification and prediction validation in complex systems. His recent interests include the sustainability of coupled interacting systems such as the SmartGrid and the interface of human and natural environments, as well as the predictability of physical behaviors exhibiting coupling between multiple underlying phenomena and scales. Dr. Ghanem has more than 100 refereed journal publications in the general areas of stochastic modeling and computations and dynamical systems. He has received several awards for his teaching and research, is the founding editor of *Lecture Notes in Mechanics* (Engineering Mechanics Institute of the American Society of Civil Engineers [ASCE-EMI]), and serves on the advisory board of a number of professional journals. He currently serves on the Board of Governors of the ASCE-EMI, is program director of the Society of Industrial and Applied Mathematics (SIAM) Activity Group on Uncertainty Quantification (SIAG/UQ), and chairs the U.S. Association for Computational Mechanics committee of SIAG/UQ.

OMAR GHATTAS is the John A. and Katherine G. Jackson Chair in Computational Geosciences and a professor of geological sciences and mechanical engineering at the University of Texas at Austin. He is also a research professor in the Institute for Geophysics, director of the Center for Computational Geosciences in the Institute for Computational Engineering and Sciences, professor of biomedical engineering and computer sciences (by courtesy), co-chief applications scientist for the 580 Teraflops NSF Track 2 supercomputer at the Texas Advanced Computing Center, and director of the KAUST-UT [King Abdullah University of Science and Technology–University of Texas at Austin] Academic Excellence Alliance. From 1989 to 2005, he was a professor at Carnegie Mellon University. He has been a visiting professor at the Institute for Computer Applications in Science and Engineering at NASA-Langley Research Center, the Center for Applied Scientific Computing at the Lawrence Livermore National Laboratory, and the Computer Science Research Institute at the Sandia National Laboratories. Professor Ghattas's research interests are in the forward and inverse modeling and the optimal design and control of complex systems in the geological, mechanical, and biomedical engineering sciences, with particular emphasis on large-scale simulation on parallel supercomputers. He received the 1998 Allen Newell Medal for Research Excellence, the Supercomputing 2002 Best Technical Paper Award, the 2003 Gordon Bell Prize for Special Accomplishment in Supercomputing, the 2004/2005 CMU College of Engineering Outstanding Research Prize, the SC2006 HPC [High Performance Computing] Analytics Challenge Award, and the TeraGrid 2008 Capability Computing Challenge Award, and he was a finalist for the 2008 Gordon Bell Prize. Professor Ghattas's recent professional activities have included service in the following capacities: he has organized 10 conferences and workshops in computational science and engineering; delivered 15 keynote or plenary talks at major international conferences; was program director for the Computational Science and Engineering Activity Group of the Society for Industrial and Applied Mathematics (SIAM); served as founding editor-in-chief of SIAM's Computational Science and Engineering series; was associate editor of *the SIAM Journal on Scientific Computing* and an editorial board member of seven other journals; served as a member of the SIAM Program Committee; and was a member of the Science Steering Committee for the Computational Infrastructure for Geodynamics project.

JUAN MEZA is dean of the School of Natural Sciences at the University of California, Merced. Prior to joining the University of California, Merced, he was the department head of High Performance Computing Research at the Lawrence Berkeley National Laboratory, where he oversaw work in computational science and mathematics, computer science and future technologies, scientific data management, visualization, and numerical algorithms and application development. His current research interests include nonlinear optimization with an emphasis on methods for parallel computing. He has also worked on various scientific and engineering applications including scalable methods for nanoscience, electric power grid reliability, cyber security, molecular conformation problems, optimal design of chemical vapor deposition furnaces, and semiconductor device modeling. Dr. Meza also held the position of Distinguished Member of the Technical Staff at the Sandia National Laboratories and served as the manager of the Computational Sciences and Mathematics Research Department. In this capacity, he acted as the Research Foundation Network Research program manager and the ASCI Problem Solving Environment Advanced Software Development Environment program manager, and he served as a member of the Sandia/California Research Council. Dr. Meza was recently named to the Top 100 Influentials list of *Hispanic Business Magazine* in the area of science. In addition, he was elected a fellow of the American Association for the Advancement of Science. In 2008, Dr. Meza was the recipient of the Blackwell-Tapia Prize and the SACNAS [Society for the Advancement of Chicanos and Native Americans in Science] Distinguished Scientist Award. He was also a member of the team that won the 2008 ACM Gordon Bell Award for Algorithm Innovation. Dr. Meza has served on numerous external committees, including the National Science Foundation's Mathematical and Physical Sciences Advisory Committee, the Department of Energy's Advanced Scientific Computing Advisory Committee, the Mathematical Sciences Research Institute's Human Resources Advisory Committee, the Board of Trustees of the Institute for Pure and Applied Mathematics, the Board of Governors of the Institute for Mathematics and Its Applications, and the Board of Trustees of the Society of Industrial and Applied Mathematics.

ERIC MICHIELSSEN is a professor of electrical and computer engineering at the University of Illinois at Urbana-Champaign (UIUC). His research interests include all aspects of theoretical and applied computational electromagnetics with an emphasis on the development of fast frequency and time domain integral-equation-based techniques for analyzing electromagnetic phenomena and robust optimizers for electromagnetic/optical devices. Professor Michielssen is the (co-)author of 120 journal articles and book chapters and 180 conference papers and abstracts. He was the recipient of a 1994 International Union of Radio Scientists (URSI) Young Scientist Fellowship, a 1995 National Science Foundation CAREER Award, and the 1998 Applied Computational Electromagnetics Society Valued Service Award. In addition, he was named the 1999 URSI United States National Committee Henry G. Booker Fellow and was selected as the recipient of the 1999 URSI Koga Gold Medal. Recently, he was awarded the UIUC's 2001 Xerox Award for Faculty Research and was appointed a Beckman Fellow in its Center for Advanced Studies, a UIUC Scholar, and a Sony Faculty Fellow. He is an associate editor for *IEEE Transactions on Antennas and Propagation* and a fellow of the Institute of Electrical and Electronics Engineers (IEEE).

VIJAYAN N. NAIR is the Donald A. Darling Professor of Statistics and a professor of industrial and operations engineering at the University of Michigan. He was chair of the Department of Statistics from 1998 to 2010. His past experience includes 15 years as a research scientist at Bell Laboratories. He has a broad range of interests in statistical methodology and applications, especially in engineering statistics. He is involved with the Center for Radiative Shock Hydrodynamics (CRASH) at the University of Michigan, one of five national centers funded under the Predictive Science Academic Alliance Program by the National Nuclear Security Administration's Office of Advanced Simulation and Computing. As part of this center, Dr. Nair has been involved in modeling and analyzing data from large-scale simulation models and in uncertainty quantification. He is the president-elect of the International Statistical Institute and president of the International Society for Business and Industrial Statistics. He is a senior fellow of the Michigan Society of Fellows and a fellow of the American Association for the Advancement of Science, the American Society for Quality, the American Statistical Association, and the Institute of Mathematical Statistics. He currently serves on the National Research Council's Board on Mathematical Sciences and Their Application, is chairing or has (co)chaired three committees, and has served on many others. He has a Ph.D. in statistics from the University of California, Berkeley.

CHARLES W. NAKHLEH manages the Inertial Confinement Fusion Target Design Department in the Pulsed Power Sciences Center at the Sandia National Laboratories. He supervises theoretical design and analysis efforts for inertial confinement fusion (ICF) targets for the Z pulsed-power facility. His department is also involved in the analysis and design of experiments for the National Ignition Campaign. Dr. Nakleh joined Sandia National Laboratories in December 2007. From 2005 to 2007, he was the group leader (acting) and deputy group leader of the Thermonuclear Applications Group (X-2) at the Los Alamos National Laboratory (LANL), where, among other tasks, he oversaw the W88 and Reliable Replacement Warhead efforts. He had spent nearly a decade before that as a staff member in X-2, working on a wide variety of weapons-physics and design issues, including the development and application of the quantification of margins and uncertainties (QMU) methodology to simulation-based predictions. Dr. Nakleh is a graduate of the Theoretical Institute of Thermonuclear and Nuclear Studies (TITANS) program at LANL. He was a member of study teams that received Department of Energy Awards of Excellence in 1999, 2000, 2005, and 2007. He has served on a wide variety of advisory panels, including as a founding member of the National Nuclear Security Administration's (NNSA's) Predictive Science Panel, the LANL director's advisory panel on weapons certification, a consultant to the 2009 JASON study on warhead Life Extension Programs, an adviser to the Undersecretary of Energy for Science on the National Ignition Campaign, and an adviser to NNSA on a variety of weapons physics issues. His research interests span a wide range of ICF, radiation effects, and other applications of high energy density physics, and applications of Bayesian inference techniques. He received his Ph.D. in physics from Cornell University in 1996.

DOUGLAS NYCHKA is the director of the Institute of Mathematics Applied to Geosciences at the National Center for Atmospheric Research (NCAR), an interdisciplinary component with a focus on transferring innovative mathematical and statistical tools to the geosciences. Dr. Nychka is a statistical scientist with an interest in the problems posed by the analysis of geophysical data sets. He received his Ph.D. from the University of Wisconsin in 1983. Subsequently he spent 14 years as a faculty member at North Carolina State University. His interest in environmental problems and a background in fitting curves and surface to spatial data led him to assume leadership of the statistics project at NCAR.

STEPHEN M. POLLOCK was the Herrick Professor of Manufacturing and a professor of industrial and operations engineering at the University of Michigan until his recent retirement. He taught courses in decision analysis, mathematical modeling, dynamic programming, and stochastic processes. His research activities include developing cost-optimal monitoring and maintenance policies, sequential hypothesis testing, modeling large multiserver systems, and dynamic optimization of radiation treatment plans. He was the director of the Program in Financial Engineering and the Engineering Global Leadership honors program. He has been an area editor of *Operations Research*, senior editor of *IIE Transactions*, president (1986) of the Operations Research Society of America, and a senior fellow of the University of Michigan Society of Fellows. Dr. Pollock is a founding fellow of the Institute for Operations Research and the Management Sciences; he was awarded its Kimball Medal in 2002. He was a member of the Army Science Board and is a member of the National Academy of Engineering. His previous National Research Council experience includes chairing the Committee on National Statistics' (CNSTAT's) Panel on Operational Test Design and Evaluation of the Interim Armored Vehicle (2002-2003), serving on the Committee on Applied and Theoretical Statistics (CATS) and on CNSTAT's Panel on Statistical Methods for Testing and Evaluating Defense Systems (1995-1998), and serving on the Committee on Modeling and Simulation for Defense Transformation and on the Committee on Methodological Improvements to the Department of Homeland Security's Biological Agent Risk Analysis.

HOWARD A. STONE is a professor in the Department of Mechanical and Aerospace Engineering at Princeton University. He received the B.S. degree in chemical engineering from the University of California at Davis in 1982 and his Ph.D. in chemical engineering from the California Institute of Technology in 1988. Following a postdoctoral year in the Department of Applied Mathematics and Theoretical Physics at the University of Cambridge, in 1989 he joined the faculty of the (now) School of Engineering and Applied Sciences at Harvard University, where he eventually became the Vicky Joseph Professor of Engineering and Applied Mathematics. In 1994 he received both the Joseph R. Levenson Memorial Award and the Phi Beta Kappa Teaching Prize, which are the only two teaching awards given to faculty in Harvard College. In 2000 he was named a Harvard College Professor for his contributions to undergraduate education. Recently he moved to Princeton University where he is the Donald R. Dixon '69 and Elizabeth W. Dixon Professor in the Department of Mechanical and Aerospace Engineering. Professor Stone's research interests are in fluid dynamics, especially as it arises in research and applications at the interfaces of engineering, chemistry, and physics. His group tackles problems with a combination of experimental, theoretical, and modeling approaches. He has received the National Science Foundation Presidential Young Investigator Award, is a fellow of the American Physical Society (APS), and is past chair of the Division of Fluid Dynamics of the APS. For 10 years he served as an associate editor of the *Journal of Fluid Mechanics*, and he is currently on the editorial or advisory boards of the *New Journal of Physics*, *Soft Matter* and *Physics of Fluids*. He is the first recipient of the G.K. Batchelor Prize in Fluid Dynamics, which was awarded in August 2008. In 2009 he was elected to the National Academy of Engineering.

ALYSON G. WILSON is a research staff member at the Institute for Defense Analyses (IDA) Science and Technology Policy Institute. Before coming to IDA, she was an associate professor in the Department of Statistics at Iowa State University. Dr. Wilson received her Ph.D. in statistics from Duke University, her M.S. in statistics from Carnegie Mellon University, and her B.A. in mathematical sciences from Rice University. She is a fellow of the American Statistical Association and a recognized expert in statistical reliability, Bayesian methods, and the application of statistics to problems in defense and national security. Prior to joining Iowa State University, Dr. Wilson was a project leader and technical lead for Department of Defense programs in the Statistical Sciences Group at the Los Alamos National Laboratory (1999-2008), a senior statistician and operations research analyst with Cowboy Programming Resources (1995-1999), and a mathematical statistician at the National Institutes of Health (1991-1992). She is a founder and past-chair of the American Statistical Association's Section on Statistics in Defense and National Security. She is a member of the *Technometrics* management committee and serves as reviews editor for the *American Statistician* and the *Journal of the American Statistical Association*. In addition to numerous publications, Dr. Wilson has co-authored a book, *Bayesian Reliability*, and has co-edited two other books, *Statistical Methods in Counterterrorism: Game Theory, Modeling, Syndromic Surveillance, and Biometric Authentication* and *Modern Statistical and Mathematical Methods in Reliability*. She holds a patent for her early work in medical imaging.

MICHAEL R. ZIKA is a project leader and an associate division leader in the AX Division at the Lawrence Livermore National Laboratory (LLNL). He earned both his B.S. (1991) and his M.S. (1992) from Purdue University, and his Ph.D. (1997) from Texas A&M University, all in nuclear engineering. In 1997 he joined LLNL as a computational physicist. His work focused on algorithms and physics models for modeling radiative transfer. As a project leader, Dr. Zika has led a large team of computational physicists and computer scientists to deliver massively parallel two-dimensional/three-dimensional multi-physics simulation tools for high energy density physics in support of the Stockpile Stewardship Program. These tools have been used to design and analyze experiments on the National Ignition Facility. In 2006 he led a team and in 2009 was a member of a team that received a Department of Energy Award of Excellence. Dr. Zika has served as an adjunct faculty member at Texas A&M University and as a visiting faculty member at the University of California, Berkeley. He has participated in a variety of strategic planning efforts at the request of the Advanced Simulation and Computing Program Office in the Department of Energy's National Nuclear Security Administration.

Appendix D

Acronyms

AD	automatic differentiation
AMR	adaptive mesh refinement
ASCE	American Society of Civil Engineers
CAD	computer-aided design
CPU	central processing unit
CRASH	Center for Radiative Shock Hydrodynamics
DEIM	discrete empirical interpolation method
DOD	Department of Defense
DOE	Department of Energy
EIM	empirical interpolation method
EMI	electromagnetic interference
GP	Gaussian process
gPC	generalized polynomial chaos
IPCC	Intergovernmental Panel on Climate Change
LANL	Los Alamos National Laboratory
LLNL	Lawrence Livermore National Laboratory
MASA	Manufactured Analytic Solution Abstraction
MC	Monte Carlo
MCMC	Markov chain Monte Carlo
MME	multimodel ensemble
MMS	method of manufactured solutions
MPS	Division of Mathematics and Physical Sciences (National Science Foundation)

APPENDIX D

M/U	margin-to-uncertainty (ratio)
NAE	National Academy of Engineering
NASA	National Aeronautics and Space Administration
NNSA	National Nuclear Security Administration
NRC	National Research Council
NSF	National Science Foundation
NWS	National Weather Service
ODE	ordinary differential equation
PC	polynomial chaos
PDE	partial differential equation
PDF	probability density function
PECOS	Center for Predictive Engineering and Computational Sciences
POD	proper orthogonal decomposition
PSAAP	Predictive Science Academic Alliance Program
QMU	quantification of margins and uncertainty
QOI	quantity of interest
SA	sensitivity analysis; Spalart-Allmaras
SC	stochastic collocation
SNL	Sandia National Laboratories
SPICE	Simulation Program with Integrated Circuit Emphasis
SQA	software quality assurance
SSP	Stockpile Stewardship Program
SUPG	streamline-upwind/Petrov-Galerkin
TPM	tire pressure monitoring
TPS	thermal protection system
UQ	uncertainty quantification
V&V	verification and validation
VVUQ	verification, validation, and uncertainty quantification